Universitext

J. Schnakenberg

Thermodynamic Network Analysis of Biological Systems

With 13 Figures

Springer-Verlag
Berlin Heidelberg New York 1977

Professor Dr. J. Schnakenberg

Institut für Theoretische Physik, Rhein.-Westf.-Techn. Hochschule
D–5100 Aachen

ISBN 3-540-08122-4 Springer-Verlag Berlin Heidelberg New York
ISBN 0-387-08122-4 Springer-Verlag New York Heidelberg Berlin

Library of Congress Cataloging in Publication Data. Schnakenberg, J. Thermodynamic network analysis of biological systems. (Universitext) Based on two lectures presented by the author at the Rheinisch-Westfälische Technische Hochschule in Aachen. Includes bibliographical references and index. 1. Biological physics. 2. Thermodynamics. I. Title. QH505.S313 574.1'912 77-1112

Printing and bookbinding: Beltz, Offsetdruck, Hemsbach
2153/3130–543210

Preface

This book is devoted to the question: What fundamental ideas and concepts can physics contribute to the analysis of complex systems like those in biology and ecology? The book originated from two lectures which I gave during the winter term 1974/75 and the summer term 1976 at the Rheinisch-Westfälische Technische Hochschule in Aachen. The wish for a lecture with this kind of subject was brought forward by students of physics as well as by those from other disciplines like biology, physiology, and engineering sciences. The students of physics were looking for ways which might lead them from their monodisciplinary studies into the interdisciplinary field between physics and life sciences. The students from the other disciplines suspected that there might be helpful physical concepts and ideas for the analysis of complex systems they ought to become acquainted with.

It is clear that a lecture or a book which tries to realize the expectations of both these groups will meet with difficulties arising from the different trainings and background knowledge of physicists and nonphysicists. For the physicists, I have tried to give a brief description of the biological aspect and significance of a problem wherever it seems necessary and appropriate and as far as a physicist like me feels authorized to do so. For the nonphysicists, the physical background information is represented in the book as far as possible, for example a brief introduction into the fundamentals of thermodynamics in Chapter 3. I would be glad if this book stimulates the physicists to further reading of physiological texts, and the nonphysicists to study a more rigorous text of irreversible thermodynamics.

The latter remark leads me to point out the significance of thermodynamics for the intention of this book. Very rigorously, one could suspect that its title is only a fashionable paraphrase of simply "the role of thermodynamics in life sciences". For two reasons, however, this would not hit the point. First, it is not just thermodynamics but thermodynamics of situations very far from equilibrium which plays the role of a ground theme for all our considerations in this book. Secondly, the conditions imposed by the fundamental laws of nonequilibrium thermodynamics will automatically be satisfied by utilizing a network language for the analysis of the systems that we have in view.

At this point, I would like to express my deep gratitude to the late Aaron Katchalsky. The lecture, "Network Thermodynamics of Membranes", which he gave on the occasion of the Israelian-German workshop in Göttingen in May 1972 and the discussions with him a few days before he was murdered on his way home to Israel represented a good deal of stimulation for me to give the lectures mentioned earlier and to write the present text.

I would also like to thank many of my colleagues in Aachen and at other universities for many helpful and elucidating discussions. In particular, I feel indebted to my coworkers Hans Josef Bebber, Christian Hook, Hans-Peter Leiseifer and Günter Wuttke who have contributed valuable results, elucidations and stimuli by their diploma theses. Thanks are expressed also to Dr. H. Lotsch (Springer Verlag) for encouraging and formulating the plan for this text. Last but not least thanks to Miss Marie-José Rozenboom for patiently, untiringly and promptly typing the manuscript.

Aachen, October 1976 J. Schnakenberg

Contents

1. Introduction

Biophysics, in its historical development as an interdisciplinary field of research between physics and biology, started from a rather particular subject, namely, the effect of irradiation, especially of high-energy X-rays, on chromosomes and the genetic information carried by them. However, biophysics has evolved to a much broader field of research encompassing a large variety of very different and vivid scientific activities between physics and biology. To give a preliminary and global definition of biophysics according to today's interpretation, one could comprise all its individual lines into the question: What methods and concepts can physics contribute to an analysis of biological systems?

Two essentially different answers may be given to this question. First, there is the physical instrumentation which is by now standard equipment of many biological or physiological laboratories. To give only a very few examples: light and electron microscopy, X-ray diffraction, EPR (electron paramagnetic resonance), NMR (nuclear magnetic resonance), high-precision measurements of electric potentials and currents including their fluctuations, and many others. Very much of special and fundamental scientific progress in biology and physiology is due exclusively to physical measurements.

In this book, however, we aim at a second answer to the above question, namely, to work out theoretical physical concepts and ideas which we hope to be appropriate for describing and explaining biological phenomena. Compared to the reality and reliability of physical instrumentation in biological and physical laboratories, our conceptual contributions to biophysics will certainly be much more speculative and controversial. Reservation against theories is already considered advisable sometimes in standard fields on monodisciplinary physical research. Consequently so much the more is this true in interdisciplinary and nonphysical fields of research. On the other hand, our relatively extensive experimental knowledge of certain biological systems makes it at least suggestive to try out theoretical physical concepts and ideas in those situations.

The way physicists introduce theoretical concepts and ideas into a phenomenologically described situation is called a "model". The method of representing an experimental result or even more than one type of result by a model is the specific concern of all of the following chapters of this book. To give a first and rough characterization, a model may be called an incomplete image of what the real system is believed to consist of, its purpose being the description or explanation of one or a few particular phenomena of the real system. In order to get acquainted with the purpose, the application, and the interpretation of models the reader will find a collection and a discussion of a few simple but typical physical models for biological systems in the first chapter of this book.

A further comment seems to be necessary concerning the phrase that the purpose of models is "to describe or to explain" an experimental phenomenon. Clearly, the most satisfactory descriptions of an experiment are the experimental curves themselves. The first step towards a model would be to reduce this usually very large amount of information by equivalently representing or simulating it in terms of as few and as simple relationships as possible, perhaps in the form of a mathematical formula. Such a reduction of information is stated by saying that a model describes an experiment. If we succeed, however, in establishing the description on the basis of the knowledge or at least of some intuitive assumptions about the specific molecular structure of the real system, we would call the model an explanation.

The use of models in the standard monodisciplinary fields of physical research is almost exclusively understood in the latter sense, namely, explaining a phenomenon by reducing it to elementary molecular structures and dynamics. For example, in solid-state physics the mechanical, thermal, and electric properties of a metal can be explained by a model in which the metal atoms are ionized and put into a perfect crystalline lattice with the electrons moving around in the lattice almost as free particles. In comparison with this relatively simple situation, a model for a semiconductor is already somewhat more complex since the electrons are no longer perfectly free and the material may be composed of more than just one type of atoms. The complexity of the model is increased even more in the case of amorphous materials, such as polymers where additionally the translational symmetry of a perfect crystal has to be replaced by some kind of a statistical disorder. We intuitively feel that with increasing complexity a model will become more and more descriptive rather than explanatory since its molecular foundation will be weakened and consequently its persuasive power of explaining a phenomenon in the literal sense. One should be very careful, however, to interpret a biological model just as quite a natural continuation of physical models like those in solid state physics towards higher degrees of complexity. First of all, biological structures, for example, a biological cell, are neither perfectly regular like crystals nor statistically disordered like gases or fluids but a very specific composition of a variety of regular structures interacting with each other in a fluid environment. Secondly, and this seems to be a consequence of the first point, our knowledge about this structure is mostly very poor. Therefore, a biological model necessarily involves a good deal of speculation and its use for describing or explaining a phenomenon always includes a considerable risk.

Realizing these latter difficulties and ambiguities when constructing models for biological systems, it would seem conclusive to refuse any kind of detailed models in biology at all. Instead, one then should consider the whole system as a black box with some external terminals being interrelated by some unknown internal processes. A number of independent experiments will in principle be sufficient to establish those interrelations at least qualitatively. The answer to the question of whether physics can also give a valuable contribution to this way of analyzing biological systems is again "yes" if one bears in mind that the black box is a system of a very large number of internal molecular degrees of freedom and thus a thermodynamic system. "Thermodynamic" means that we are not interested in the individual dynamics of the molecular degrees of freedom but only in a much smaller number of macroscopic processes which are represented by the external terminals. This in turn implies that the processes which we observe at the external terminals are to be submitted to the fundamental laws of thermodynamics, i.e., to the first law expressing the conservation of energy and to the second law expressing the increase of entropy. (The third law of the inaccessibility of the absolute zero of temperature is evidently irrelevant for biological systems). Indeed, one can derive very general conclusions from the thermodynamic laws for every particular black box, but just because of this generality the nature of such thermodynamic conclusions will always be such that certain processes at the external terminals will never be observed. A definite prediction, on the other hand, of what actually should be expected at the terminals under given external conditions can only be obtained from model studies but never from thermodynamics. In the second chapter of this book, the reader will find a brief introduction to thermodynamics and learn how such restrictive conclusions are derived from its first and second law. If he feels sufficiently acquainted with that, he may of course skip this chapter.

What physics actually can contribute after all to the analysis of biological systems is a combination of model studies together with a thermodynamic treatment of a biological system. The best way to accomplish this combination is to utilize a model language which identically satisfies the thermodynamic laws. Such a language has been proposed by OSTER, PERELSON and KATCHALSKY in 1973: network thermodynamics. Being a network language, it also automatically accounts for the discontinuous structure of biological systems. We shall extensively utilize the language of network thermodynamics and therefore describe it in Chapter 4. All that follows then in the remaining chapters of the book may be characterized as an attempt to translate already accepted physical models for biological systems into the network language, to formulate new models in this language, and to survey these models with regard to their essential mathematical and physical properties on the basis of a network representation.

2. Models

2.1 The Purpose of Nature of Models

A model always means a more or less drastic simplification of some real system in order to describe or even to explain a phenomenon which is experimentally observed in that real system. A description of a phenomenon is a mathematical formulation of its constitutive variables, mostly in the form of a differential equation or a system of differential equations which contain the experimental parameters of the phenomenon as constants. An explanation goes beyond a pure description in that it reduces the occurrence of the phenomenon to more elementary relationships or laws which were known already. It is evident that the border line between a descriptive and an explanatory model is hazy and very often depends on the point of view.

The purpose of the simplification process in constructing a model is to reveal the mechanisms and interactions which are believed or even known to be responsible for that particular phenomenon. All details which do not directly influence the phenomenon or modify it only quantitatively but not qualitatively will be neglected, at least in the ground version of the model. Thus, a model represents a very incomplete picture of the real system. It may even happen that two models for two different phenomena which are observed in the same real system have none of their elements in common. Only a combination of the various models for phenomena in the same real system may possibly lead to something like a complete picture if such a combination does not meet with difficulties due to mutual incompatibility of the different models.

In the second chapter of this book, we shall represent and discuss a few examples of physical or chemical models for biological phenomena like transport across membranes, membrane excitation, control of metabolism, and population dynamic interaction between different species. All these models will be of the type of a reaction kinetic model, i.e., the model processes are chemical reactions and diffusion of molecules or may at least be interpreted like that. Thus, the physical background of the various models is irreversible thermodynamics of reactions and diffusion. This background will be briefly reviewed in the next chapter. The common physical background behind our models will then lead us in the following chapters to a generalized language for the representation of the models in terms of networks. We shall learn to treat this language and to apply it to the particular models represented in this chapter as well as to make use of quite general physical and mathematical network techniques to predict the qualitative properties of a model on the basis of its topological network structure.

Being a reaction kinetic model still does not sufficiently characterize the type of a model. A reaction kinetic model may be as detailed as to display each of the single molecular steps of the reactions and the diffusion processes from which the phenomenon in view is believed to be composed; but it also may integrate the single steps into a smaller number of effective or over-all reactions and diffusion processes. In the first case, one would tend to call the model a molecular model, in the latter case a phenomenological or black-box model. But even in the case of a molecular model in this sense, the model description is still very far from what is usually understood as a microscopic description in physics. The dynamical basis of microscopic models is quantum mechanical equations of motion which are reversible with respect to inversion of time. A single molecular reaction or diffusion step, however, is already the result of an elimination of a large number of reversible microscopic degrees of freedom. This is reflected by the fact that the equations of motion for reactions and diffusion are irreversible with respect to inversion of time which makes them a subject of irreversible thermodynamics rather than of quantum mechanics.

Closely related to the question of how detailed a model claims to display the molecular structure of the real system is the number of parameters of the model. It is evident that the more details are included in the model the larger this number becomes. On the other hand, if only a limited amount of experimental data is available from the real system, a model with a large number of parameters does not seem to make much sense since the adjustment of the parameters to the experimental data will then be a rather ambiguous procedure. This is a very crucial point when designing physical models for biological systems. Whereas in purely physical fields like high energy physics, nuclear physics or solid state physics there is relatively much and ensured information regarding the nature of the underlying interactions and the symmetry of the real system, this information is mostly very poor in biological systems. Also, the interpretation of the experimental results is much more difficult in biological system than in purely physical systems since the high complexity of composition in biological systems makes it very ambiguous to correlate experimental data with particular molecular interactions.

On these grounds, phenomenological models for biological systems will usually be more realistic than those involving a large number of assumptions on the molecular level. We shall see that the network language represents itself as a reasonable basis for reducing a model to its skeleton structure or minimal version which is essential for the description of a certain phenomenon. This language may also serve to compare different models of the same phenomenon with respect to their essential characters and thus to decide whether they are really different or equivalent from the structural point of view.

2.2 Enzyme-Catalyzed Reactions and the Michaelis-Menten Kinetics

Most of the metabolic reactions in living organisms would not proceed at a physiologically reasonable rate if specific enzymes which act as catalysts were not present. Enzymes are special protein molecules, i.e., biopolymers with amino acids as monomeric units of a polypeptide chain and with a molecular weight of the order up to 10^5. Since there are 20 different amino acids, a tremendous number of different proteins could be formed among which, however, only a very small fraction is stable and biologically meaningful. For the purpose of this book, we need not go into the details of protein chemistry which may be found in any textbook of biochemistry. (For a brief abstract see CHAPMAN, LESLIE (1967)). The only point that we adopt from enzyme chemistry is the assumption of a specific stereochemical configuration or even a binding site which is formed by the spatial structure of the polypeptide chain of the enzyme E and selectively offered to a substrate S to be transferred to some product P by the metabolic reaction. The reaction from S to P is assumed to be performed while S is bound to its enzyme E. The reaction scheme of this process is then simply given by

$$E + S \rightleftharpoons [ES] \rightleftharpoons E + P \tag{2.1}$$

where [ES] denotes the bound enzyme-substrate complex. To obtain a quantitative description for our reaction system, we introduce the reaction rates J_1 and J_2, i.e., the number of moles per second transferred in the two reaction steps of (2.1). Denoting the concentrations (mole per liter) of free and bound enzyme by the symbols E and [ES] we clearly have

$$\frac{d}{dt} E = - J_1 + J_2 \ , \qquad \frac{d}{dt} [ES] = J_1 - J_2 \tag{2.2}$$

for the time changes of the concentrations. A particular consequence of (2.2) is

$$\frac{d}{dt} (E + [ES]) = 0 \ , \qquad E + [ES] = E_0 = \text{const} \ , \tag{2.3}$$

expressing the fact that the total amount of the enzyme as a catalyst is conserved.

The next step in the development of our simple model is an ansatz for the reaction rates J_1, J_2 in terms of the concentrations of the chemical species involved in these reactions:

$$\left.\begin{aligned} J_1 &= k_1 SE - k_1'[ES] \\ J_2 &= k_2[ES] - k_2'PE \end{aligned}\right\} . \tag{2.4}$$

Eq. (2.4) says that the reaction rates can be written as a difference or balance of a forward and a reverse rate each of which is proportional to the products of the involved concentrations. The proportionality constants are called the reaction rate constants. Although the ansatz (2.4) seems to be very intuitive, it is far from being trivial even for molecules with a much less complicated structure than that of enzymes and metabolic substrates and products. For the justification of (2.4) for enzyme reactions, the reader is referred to the special literature, for example FORST (1973), ROBINSON and HOLBROOK (1972), LAIDLER and BUNTING (1973).

Let us now study the steady state properties of our reaction system. To this purpose, we consider the concentrations of substrate S and product P as kept fixed by some appropriate transport process. The steady state is then defined as a state where all time derivatives vanish. We expect that such a state will be achieved for $t \longrightarrow \infty$ provided that all parameters are kept constant. From (2.2) we see that the steady state in our system implies $\bar{J}_1 = \bar{J}_2$, the bar indicating the steady state condition. Inserting the ansatz (2.4) we obtain

$$\overline{[ES]} = \frac{k_1 S + k_2' P}{k'_1 + k_2} \bar{E} \quad . \tag{2.5}$$

By making use of $E + [ES] = E_0$, (2.3), we can evaluate the steady state concentrations $\overline{[ES]}$ and $\bar{E}$ from (2.5) in terms of S, P and E_0. Substituting the result back into (2.4) finally yields

$$\bar{J} = \bar{J}_1 = \bar{J}_2 = \frac{k_1 k_2 S - k_1' k_2' P}{k_1' + k_2 + k_1 S + k_2' P} E_0 \quad . \tag{2.6}$$

For an application of (2.6) to enzyme experiments, one usually assumes $P \approx 0$, which means an elimination of P from the reaction by a very fast transport process, or $k_2' \approx 0$, which means that the binding site of the enzyme has a high attractivity only to the substrate S but not to the product P. With this assumption, the plot of $\bar{J}$ versus S has the form given in Fig. 1 (Michaelis-Menten kinetics)

For high values of S, the reaction rate $\bar{J}$ shows saturation. A comparison of the experimental plot of $\bar{J}$ versus S allows a determination of the rate constants.

Let us conclude the discussion of our simple model by an inspection of the time-dependent properties. It is elucidating for the physical structure of the model to introduce probabilities p_0 and p_1 for an enzyme molecule in its free and its bound form E and [ES], respectively:

$$p_0 = E/E_0 \ , \qquad p_1 = [ES]/E_0 \tag{2.7}$$

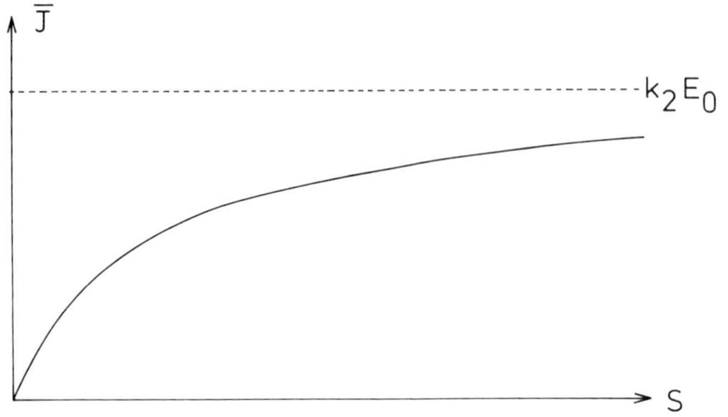

Fig. 1. Michaelis-Menten behaviour of the steady state flux $\bar{J}$ as a function of the substrate concentration S

such that

$$p_o + p_1 = 1 \ . \qquad (2.8)$$

From (2.2) and (2.4) we easily derive an expression for the time-derivative of p_o and p_1:

$$\frac{dp_o}{dt} = (k_1' + k_2)p_1 - (k_1S + k_2'P)p_o \quad . \qquad (2.9)$$

Due to the normalization condition (2.8), the expression for dp_1/dt is equal to $-dp_o/dt$. Eq. (2.9) is a particularly simple example for a linear rate equation the general form of which reads

$$\frac{dp_i}{dt} = \sum_j (U_{ij}p_j - U_{ji}p_i) \qquad (2.10)$$

where $\{p_o, p_1, \dots p_N\}$ is a set of probabilities for the case that the system has N + 1 states. In the physical literature, an equation of the type of (2.10) is generally called a "master equation".

Setting $dp_o/dt = 0$ in (2.9) we can again immediately calculate the steady state probabilities $\bar{p}_o = \bar{E}/E_o$ and $\bar{p}_1 = \overline{[ES]}/E_o$ by making use of the normalization condition (2.9). Let us now denote the deviations of p_o and p_1 from the steady state values by

$$\delta p_o(t) = p_o(t) - \bar{p}_o \ , \qquad \delta p_1(t) = p_1(t) - \bar{p}_1 \quad . \qquad (2.11)$$

From (2.8) it follows that

$$\delta p_0(t) + \delta p_1(t) = 0 \ . \tag{2.12}$$

Taking the time-derative of $\delta p_0(t)$ we obtain from (2.9) together with (2.12)

$$\frac{d}{dt}\,\delta p_0 = (k_1' + k_2)\delta p_1 - (k_1 S + k_2' P)\delta p_0 = -\,(k_1' + k_2 + k_1 S + k_2' P)\delta p_0 \ . \tag{2.13}$$

The solution of (2.13) is

$$\delta p_0(t) = \delta p_0(0) \cdot e^{-t/\tau} \tag{2.14}$$

with

$$\frac{1}{\tau} = k_1' + k_2 + k_1 S + k_2' P \ . \tag{2.15}$$

Eq. (2.14) means that the probabilities decay exponentially to their steady state values. This phenomenon is called a relaxation process with τ as the relaxation time. In principle, relaxation can be observed experimentally after an instantaneous change of the values of the external parameters S and P. If such an experiment is practicable, the plot of $1/\tau$ versus S would yield information about the rate constants independently of the steady state measurements.

2.3 Transport Across Membranes: A Black-Box Approach

Biological membranes are the "wall elements" in the organization of living organisms. Every cell is surrounded by a membrane and thus receives its individuality from the membrane. The same applies to the sub-units of the cell as for example to the nucleus or the mitochondria (the power supply of the cell). Being a wall, however, is only one aspect of biological membranes and certainly not the most important one since a completely separating wall around a cell would make the cell senseless for the rest of the living organism. The biologically much more relevant function of membranes is that of a contact element for example between the cell and its surrounding. This contact function mainly consists of transport of molecules or ions across the membrane. The transported molecules may also carry energy or information into or out of the cell.

As a first step towards a model for transport across membranes, let us consider a very simple situation where just one kind of a neutral molecule is transported across the membrane under the action of its own concentration difference $\Delta c = c - c'$, c and c' being the concentrations of the molecule at the two sides of the membrane. The simplest ansatz for the flux J of the molecule, i.e., the number of moles penetrating through the membrane per time and area, would be Fick's first law,

$$J = D \frac{\Delta c}{a} \qquad (2.16)$$

where a ist the thickness of the membrane, which is of the order of 100 Å for biological membranes, and D is an effective diffusion coefficient.

The ansatz (2.16) is a typical black-box approach in that it completely neglects all details of membrane structure and molecular transport mechanisms. Let us represent this ansatz by a very simple network where the transport process is expressed as a material resistance R upon which the chemical potentials μ and μ' of the molecule at the two sides of the membrane are acting:

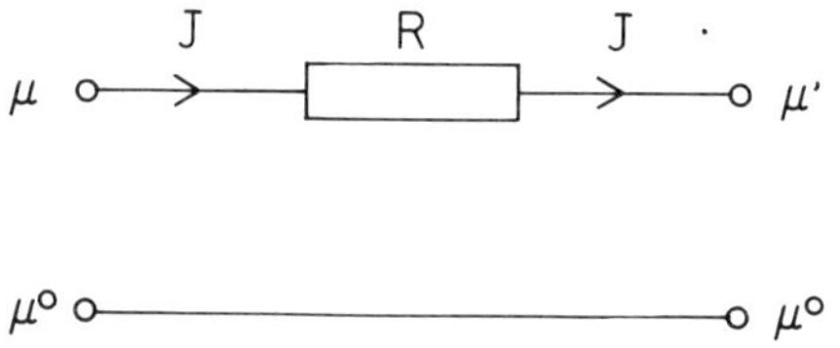

Fig. 2. Representative network for material transport across a membrane, μ^0 being an arbitrary reference or ground potential

For this network, we can formulate a relation which is analogous to Ohm's law in an electric network, namely

$$J = \frac{1}{R}(\mu - \mu') = \frac{1}{R}\Delta\mu \qquad (2.17)$$

where the chemical potentials μ, μ' simply replace the electric potentials in the electric case. Clearly, (2.16) and (2.17) express the same physical content. This means that there must exist a relationship between the diffusion constant D and the material resistance R. This relationship is easily established if we make use of the relation between the chemical potential μ and its concentration c in dilute solutions:

$$\mu = \mu^0 + RT \ln c \qquad (2.18)$$

and similarly for μ' and c'. μ^0 is the so-called reference potential in the solution which, for the sake of simplicity, we also could have chosen as the potential of the ground line in our network in Fig. 2. The important point is that μ^0 depends on the thermodynamic parameters of the solution but not on the concentration c. $\mathcal{R}$ is the gas constant and T is the absolute temperature. From (2.18) we obtain c as a function of its potential μ. Inserting this relation into (2.16) and noticing that the reference potential μ^o is the same on both sides of the membrane, we get

$$J = \frac{D}{a}\exp\left(-\frac{\mu^o}{\mathcal{R}T}\right)\left(\exp\frac{\mu}{\mathcal{R}T} - \exp\frac{\mu'}{\mathcal{R}T}\right) = \frac{D}{a}\left(\exp\frac{\bar{\mu}-\mu^o}{\mathcal{R}T}\right)\left[\exp\left(\frac{\Delta\mu}{2\mathcal{R}T}\right) - \exp\left(-\frac{\Delta\mu}{2\mathcal{R}T}\right)\right]$$
$$\approx \frac{D\bar{c}}{a\mathcal{R}T}\Delta\mu \qquad (2.19)$$

for small values of $\Delta\mu$. $\bar{c}$ is an average concentration defined as

$$\bar{c} = \exp\left(\frac{\bar{\mu}-\mu^o}{\mathcal{R}T}\right), \qquad \bar{\mu} = \frac{1}{2}(\mu + \mu') \; . \qquad (2.20)$$

Comparison of (2.19) with (2.17) yields

$$R = \frac{a \cdot \mathcal{R}T}{D \cdot \bar{c}} \qquad (2.21)$$

for the relation between R and D.

So far, our simple model only describes steady state transport across membranes, but no relaxation. In the network of Fig. 2, the flux J would adjust to a new steady state value without time delay after a change of the potential difference. In order to exhibit relaxation behaviour, the membrane must be capable of storing the molecules which are transported across the membrane. The network representation of such a storage phenomenon is a material capacitance C_m which has to be added to the network as shown in Fig. 3:

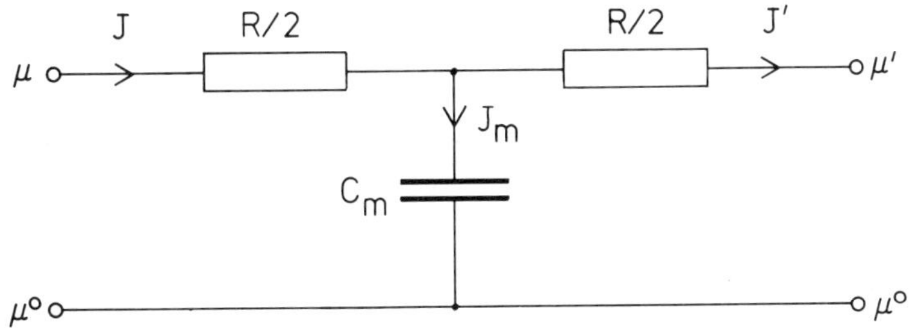

Fig. 3. Network for material transport across membranes including relaxation

In Fig. 3, we have assumed that the membrane is symmetric such that the material resistance is split into two equal parts R/2 between which the capacitance C_m has been placed. The definition of the material capacitance follows that of an electrical capacitance, namely

$$J_m = C_m \frac{d\mu_m}{dt} \tag{2.22}$$

where J_m is the material flux into C_m and μ_m is the chemical potential of the molecule within the membrane. Assuming the same dependence of μ_m on the concentration c_m in the membrane interior as in the solution,

$$\mu_m = \mu_m^o + RT \ln c_m \tag{2.23}$$

where the membrane reference potential μ_m^o may be different from that in the solution, we obtain

$$\frac{d\mu_m}{dt} = \frac{RT}{c_m} \frac{dc_m}{dt} \quad . \tag{2.24}$$

Since $J_m = a dc_m/dt$ (a = membrane thickness) we conclude from (2.22) and (2.24)

$$C_m = \frac{c_m a}{RT} = \frac{a}{RT} \exp\left(\frac{\mu_m - \mu_m^o}{RT}\right) \quad . \tag{2.25}$$

In contrast to the electric capacitance, the material capacitance C_m is not a constant but varies with the membrane concentration c_m or its chemical potential μ_m.

What remains to be done to obtain the relaxation behaviour is to apply Kirchhoff's current law,

$$J_m = J - J' \tag{2.26}$$

and to insert J_m from (2.22) and J, J' from Ohm's law for both parts of the material resistance:

$$J = \frac{1}{R/2} (\mu - \mu_m) \quad , \quad J' = \frac{1}{R/2} (\mu_m - \mu') \tag{2.27}$$

which yields

$$C_m \frac{d\mu_m}{dt} = - \frac{4}{R} \mu_m + \frac{2}{R} (\mu + \mu') \quad . \tag{2.28}$$

From (2.28) we deduce for the steady state, $d\mu_m/dt = 0$, $\mu_m = (\mu + \mu')/2$ which upon insertion into (2.27) leads back to (2.17).

The calculation of the relaxation behaviour requires the solution of a nonlinear differential equation (2.28), the nonlinearity being caused by the dependence of C_m on the variable μ_m. In order to avoid this mathematical difficulty, let us give an approximate solution by assuming that the time variations of μ_m are so small that C_m may be considered as constant. With this approximation, we obtain from (2.28) for the deviation $\delta\mu_m = \mu_m - \bar{\mu}_m$ of μ_m from its steady state value $\bar{\mu}_m$ a linear differential equation

$$\frac{d}{dt}\,\delta\mu_m = -\frac{4}{RC_m}\,\delta\mu_m \tag{2.29}$$

the solution of which reads

$$\delta\mu_m(t) = \delta\mu_m(0)e^{-t/\tau} \tag{2.30}$$

with the relaxation time τ given by

$$\tau = \frac{1}{4}\,RC_m\ . \tag{2.31}$$

Inserting now the expressions (2.21) and (2.25) for the material resistance R and capacitance C_m into (2.31) and assuming that the mean concentrations $\bar{c}$ of (2.21) and $\bar{c}_m$ of (2.25) are the same, we obtain a relation which can be written as

$$2\,\frac{D}{2}\,\tau = \left(\frac{a}{2}\right)^2\ . \tag{2.32}$$

This is nothing else than Einstein's diffusion relation for diffusion across a symmetric membrane, the factors 1/2 being due to the splitting of the membrane into two symmetric parts.

The black-box network approach of this section is taken from a review article of OSTER, PERELSON and KATCHALSKY (1973), which is recommended for further details and the physical background of this approach. A general introduction into the network language will be given in Chapter 4 of this book.

2.4 Excitation of the Nervous Membrane: Hodgkin-Huxley Equations

The nervous membrane is the wall of the nervous cell. Its excitation is the basic process of the propagation of information along the axons, which are the conduction elements of the nervous cell. The essential variables of the excitation process are the ionic fluxes of Na^+ and K^+ across the membrane and the electric membrane voltage $V = \Phi_a - \Phi_e$ where Φ_a and Φ_e are the electric potentials of the axoplasm, the interior of the nerve cell, and of the external medium of the cell, respectively. A local record of the membrane voltage V while the axon is transmitting a signal shows a train of voltage peaks with an amplitude of the order of 100 mV above a ground level of the order of -60 mV, cf. Fig. 4.

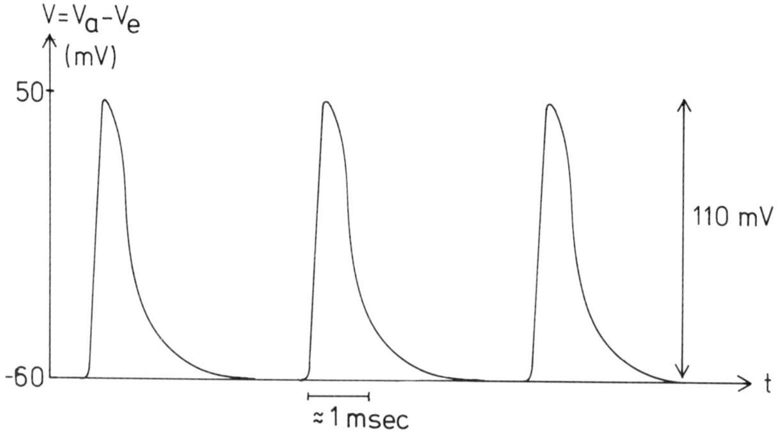

Fig. 4. Series of action potentials during transmittance of nervous information

These voltage peaks or action potentials travel along the axon and carry the information with them. The intensity of information is coded as the frequency or "firing rate" of the action potentials. This frequency usually varies between 5 and 100 Hz; maximum frequencies have been observed up to 1,000 Hz. This implies a time interval of a few microseconds for the single action potentials.

Since the work of BERNSTEIN (1902, 1912) it is known that nervous conduction is a membrane phenomenon. A systematic description of the excitation of the giant axon of squid has been given by HODGKIN and HUXLEY (1952). The diameter of about 1 mm of this axon makes it a favourite candidate for laboratory investigations. For a detailed presentation of this description the reader is referred to a review lecture of HODGKIN (1967) or to the monograph of COLE (1968); for a general introduction into the physiology of the nervous system, a textbook on neurophysiology like that of SCHMIDT (1975) is recommended.

The basis of the Hodgkin-Huxley descriptive model is the qualitative resemblance of the resting potential $V = V_r \approx -60$ mV between the excitation peaks and of the peak voltage $V = V_p \approx +50$ mV with the Nernst potentials of K^+ and Na^+ across the nervous membrane:

$$\left.\begin{aligned} V_K &= \frac{RT}{F} \ln \frac{[K^+]_e}{[K^+]_a} \approx -75 \text{ mV} \\ V_{Na} &= \frac{RT}{F} \ln \frac{[Na^+]_e}{[Na^+]_a} \approx +55 \text{ mV} \end{aligned}\right\} \tag{2.33}$$

where $[K^+]_a$, $[K^+]_e$, $[Na^+]_a$, $[Na^+]_e$ are the axoplasm and external concentrations of K^+ and Na^+, respectively, and F is Faraday's constant. The voltage values given in (2.33) imply $[K^+]_a > [K^+]_e$ and $[Na^+]_a < [Na^+]_e$. The fact that $V_r \approx V_K$ and $V_p \approx V_{Na}$ suggests that the membrane is K^+-permeable and Na^+-impermeable in the rest state and vice versa in the excited state. This suggestion is confirmed by tracer experiments during excitation showing a Na^+-conduction peak almost parallel to and as pronounced as the V-peak. This Na^+-conduction peak is followed by a time-delayed and much softer K^+-conduction maximum which restores V to its resting value.

The whole excitation phenomenon travels along the axon membrane. Within the framework of our brief representation in this section, we restrict ourselves to a local description of the phenomenon which has been given by Hodgkin and Huxley in the form of a network shown in Fig. 5.

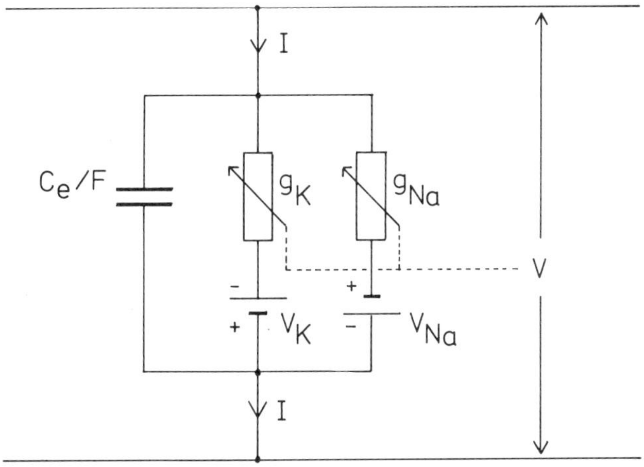

Fig. 5. Hodgkin-Huxley network for membrane excitation

In Fig. 5, the batteries V_K, V_{Na} represent the Nernst potentials, C_e/F is the electrical capacitance per area between the adjacent axoplasm and external medium and the material resistance elements with conductivities g_K and g_{Na} stand for the K^+- and Na^+-conduction channels across the membrane. The total electric flux I is then given by

$$I = \frac{C_e}{F}\frac{dV}{dt} + g_K \cdot (V - V_K) + g_{Na} \cdot (V - V_{Na}) \ . \tag{2.34}$$

The important point of this model description is the appearance of a feedback coupling from the membrane voltage V to the conductivities g_K, g_{Na} indicated as dotted lines in Fig. 5. If such a feedback coupling were not present, the system would behave like an ordinary electric network and relax exponentially into its steady state but never show excitability. Hodgkin and Huxley have shown that this necessary feedback coupling can be represented by three different kinds of charged particles which control the values of g_K and g_{Na}. Let p_1, p_2 and p_3 the probabilities of the particles to be present in the membrane at the positions of the channels. It is found now that the K^+-channel is activated by the first particle in a fourth-order chemical reaction but independent on the second and third particle, whereas the Na^+-channel is independent on the first particle but activated by the second particle in a trimolecular reaction and inhibited by the third particle in a unimolecular reaction such that

$$g_K = g_K^o \cdot p_1^4 \quad , \qquad g_{Na} = g_{Na}^o \cdot p_2^3 \cdot (1 - p_3) \tag{2.35}$$

where g_K^o and g_{Na}^o are constants. For each of the p_i, $i = 1,2,3$, a master equation like that in (2.9) or (2.10) is formulated. Taking into account only two states for each particle, namely that of being present with probability p_i and that of being not present with probability $1 - p_i$, one thus obtains

$$\frac{dp_i}{dt} = k_i\,(1 - p_i) - k_i' p_i \qquad i = 1,2,3 \quad . \tag{2.36}$$

Since the transitions described in (2.36) are associated with a change of position of charged particles, the rate constants k_i and k_i' will depend on the membrane voltage V which establishes the feedback coupling. For the appearance of excitation, this feedback coupling must be positive as the reader may easily conclude from Fig. 5 by qualitative arguments. In the following section and in Section 6.2, we shall represent two more detailed excitation models which explicitly show this sign condition of feedback. As a consequence, we must have

$$\frac{dk_i}{dV} > 0 \; , \quad \frac{dk_i'}{dV} < 0 \quad . \tag{2.37}$$

With an appropriate choice of the V-dependence of k_i and k_i' the model is complete and may be evaluated on a computer. It turns out that the parameters of the model can be adjusted in such a way that the computed curves fit the experimental ones sufficiently well. We do not want to go into any details of this evaluation but once more stress the feedback character of the model. This latter point will lead us to quite different excitation models, namely, a molecular model in the following section and a dissipative one in Section 6.2. We will come back to the Hodgkin-Huxley model in Section 5.6 in context with a generalized network language.

2.5 A Cooperative Model for Nerve Excitation

The discussion of a positive feedback as the deciding point of excitation models leads us directly to a class of models in which a cooperative mechanism from statistical physics is applied to nerve excitation. This cooperative mechanism is the Ising model; it is frequently used to describe phase transitions in solids and liquids and its application to nerve excitation is very suggestive. The first one who worked out an excitation theory on the basis of the Ising model was ADAM (1968; 1970). Meanwhile, a number of variations of this idea has been proposed by BLUMENTHAL, CHANGEUX, LEFEVER (1970), HILL, CHEN (1971), BASS, MOORE (1973), KARREMANN (1973), GOTOH (1975). Since the basic idea is the same in all these theories, let us restrict ourselves to a brief outline of Adam's model.

In Adam's model, it is assumed that the axon membrane contains a more or less irregular lattice of so-called active centers. Each of the active centers is supposed to bind either a bi-valent Ca^{2+} ion or two mono-valent K^+ ions. The state with Ca^{2+} bound to the active center is called the ground state, that with two mono-valent K^+ ions the excited state. An active center may pass from the ground state to the excited state by releasing the Ca^{2+} ion into the axoplasm or into the exterior and by taking up two K^+ ions from the axoplasm or from the exterior, and quite similarly for the reverse transition. Let p be the probability for an active center to be in the excited state and hence 1-p that for the ground state. For the time behaviour of p, a master equation ansatz is made like that in the Hodgkin-Huxley theory, cf. (2.36)

$$\frac{dp}{dt} = k\,(1 - p) - k'p \quad . \tag{2.38}$$

The transition probabilities or rate constants k and k' are functions of the concentrations of Ca^{2+} and K^+ on both sides of the membrane and of the membrane voltage V. Let c_{1a} and c_{1e} be the concentrations of K^+ in the axoplasm and in the exterior solution, respectively, and similarly c_{2a} and c_{2e} that of Ca^{2+}. The explicit dependence of k and k' on c_{1a}, c_{1e}, c_{2a}, c_{2e} and V is then given by

$$\left.\begin{aligned} k &= \alpha c_{1a}^2 e^{2\varphi} + \beta c_{1a} c_{1e} + \gamma c_{1e}^2 e^{-2\varphi} \\ k' &= \kappa c_{2a} e^{2\varphi} + \lambda c_{2e} e^{-2\varphi} \end{aligned}\right\} \tag{2.39}$$

where

$$\varphi = \frac{FV}{2RT} \quad . \tag{2.40}$$

The constants $\alpha,\beta,\gamma,\kappa,\lambda$ include relative weighting factors expressing that the ions can be taken up either from the axoplasm or from the exterior of the axon. The exponential form of the dependence on the membrane voltage V takes into account that in the case of ions the chemical potential μ has to be replaced by the electrochemical potential $\eta = \mu + zF\Phi$ where z is the valency of the ion ($z = +1$ for K^+ and $z = +2$ for Ca^{2+}) and Φ is the electrical potential at the position of the ion. Remembering (2.18) for the relation between the chemical potential and concentration of uncharged molecules, we have to replace the effective concentration factors for ions by

$$c \longrightarrow \exp\frac{\eta - \mu^0}{RT} = c \exp\frac{zF\Phi}{RT} \quad . \tag{2.41}$$

Assuming now that the binding sites of the active centers have positions precisely in the center of the membrane and choosing the electric potentials in the axoplasm and in the exterior as $\Phi_a = V/2$, $\Phi_e = -V/2$, we can explicitly derive the expressions (2.39). We shall come back to this point in Chapter 3.

So far, the model does not yet contain any kind of a cooperative or feedback mechanism. Feedback is brought into play now by the assumption that the active centers are not independent of each other but that the transition probability for a certain active center to change from the ground to the excited state or vice versa depends on the states of the neighbouring centers at the time of the transition. In the Ising model, this interaction between neighbouring centers is expressed by interaction energies w_0 and w_1 such that a Ca^{2+} ion bound to a certain center in the ground state receives an additional binding energy w_0 for each neighbouring center which also has bound Ca^{2+}. Similarly, a pair of K^+ ions bound to an excited center receives an additional binding energy w_1 for each neighbouring center which also is in the excited state. It is obvious that this kind of interaction is cooperative in the sense that it causes a self-stabilization of clusters of active centers which are in the same state.

The precise formulation of the interaction mechanism of the Ising model requires one to introduce separate probabilities p_i governed by separate master equations of the type of (2.38) for each individual active center. At this point, we introduce an approximation which is known in statistical physics as mean field or Bragg-Williams approximation. It takes into account the interaction between active centers by making the rate constants k and k' for the average probability p in (2.38) functions of p such that

$$k = k_o \exp\left[-\frac{\nu w_o}{RT}(1 - p)\right] \quad , \quad k' = k'_o \exp\left(-\frac{\nu w_1}{RT}\right) \quad . \tag{2.42}$$

In (2.42), we recognize the expressions $\nu w_o(1 - p)$ and $\nu w_1 p$ as the average interaction energies which receive an active center in the ground and the excited state from their nearest neighbours if ν is the number of nearest neighbours. The particular form of the dependence of k and k' on these average interaction energies follows from the fact that a rate constant k always shows a so-called activation energy factor of the Arrhenius type exp (- A/RT) where A is the energy threshold for the corresponding reaction step. It is clear that the average interaction between the active centers gives an additive contribution to A which directly leads to (2.42).

Let us evaluate now (2.38) together with (2.42) in the steady state, i.e., dp/dt = 0, yielding

$$\frac{\bar{p}}{1 - \bar{p}} \exp\left(-\alpha\bar{p}\right) = \frac{k_o}{k'_o} \exp\left(-\frac{\nu w_o}{RT}\right) \equiv f(V) \tag{2.43}$$

where $\alpha = \nu(w_o + w_1)/RT$ and for k_o and k'_o the right-hand sides of (2.39) have to be inserted. This makes the right-hand side of (2.43) a function of the membrane voltage V and all ionic concentrations. Regarding the V-dependence, we notice that there is almost no Ca^{2+} present in the axoplasm, i.e., $C_{2a} \approx 0$, such that k_o/k'_o and thus f(V) in (2.43) becomes a monotonically increasing function of V. The steady state solutions $\bar{p}$ may be obtained from (2.43) by plotting $\bar{p}$ as a function of $\xi = \ln f(V)$ which is shown in Fig. 6.

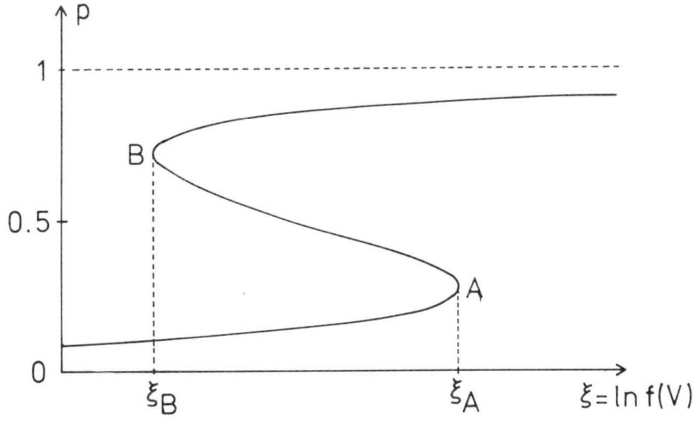

Fig. 6. Probability p of an excited active center as a function of $\xi = \ln f(V)$ in mean field approximation ($\alpha > 4$)

For $\alpha > 4$, there exists an interval $\xi_B < \xi < \xi_A$ where $\bar{p}$ as a function of ξ is triple-valued, whereas for $\alpha < 4$ $\bar{p}$ is always single-valued. Let V_A and V_B be the values of the membrane potential corresponding to the turning points ξ_A and ξ_B.

Then for membrane potentials V within the interval $V_B < V < V_A$, there appear three different steady states of the membrane. By means of a relaxation analysis of (2.38) like that presented in Sections 2.2 and 2.3 one can show, however, that the central branch between the turning points A and B in Fig. 6 is instable. Thus, only the lower and the upper branches represent stable steady states of the system.

The cooperative excitation model is completed now by the assumption that the collective states of the membrane control the Na^+-conductivity g_{Na} such that g_{Na} is almost zero in the ground state and finite in the excited state. The transport of K^+ is supposed to be carried by separate channels such that the K^+-conductivity g_K is independent of the state of the membrane. If now the resting potential V_r is smaller than V_B, the average probability p will be almost zero in the resting state which means that almost all centers are in the ground state and the membrane is Na^+-impermeable. As V is increased in the course of an action potential peak the steady state will remain on the lower branch with small values of p until V passes the turning point V_A at which a discontinuous transition to the upper branch with values of p almost equal to one takes place. The membrane is now largely excited and Na^+-permeable. In the opposite direction, the steady state will follow the upper branch until V is decreased below V_B and then jump back into the lower branch. The complete cycle of an action potential thus shows a hysteresis behaviour like that of the magnetization in a ferromagnet as a function of the external magnetic field.

Like the brief review of the Hodgkin-Huxley model in the preceding section, the cooperative model presented in this section is a local excitation model and does not include the propagation mechanism of excitation. It does not tell us why the local value of V is increased while an action potential peak is travelling along the axon, but only how the local membrane properties and the Na^+-conductivity are discontinuously changing during an action potential. Let us once more repeat that the essential point of the above model is the existence of two stable steady states due to the positive feedback which manifests itself as a cooperative interaction with positive interaction energies between the neighbouring active centers. The reader may verify that under certain conditions the positive feedback in the Hodgkin-Huxley model also produces two stable steady state, cf. problem 3. In this sense, Adam's model can be considered as a molecular realization of the general idea behind the Hodgkin-Huxley theory.

2.6 The Volterra-Lotka Model

The model which we shall discuss in the present section is concerned with an ecological problem. Almost 50 years ago, VOLTERRA (1928, 1931) set up a system of differential equations for the interaction of a prey species X and a predator species Y:

$$\left.\begin{aligned}\frac{dX}{dt} &= k_1X - k_3XY \\ \frac{dY}{dt} &= -k_2Y + k_4XY\end{aligned}\right\} . \tag{2.44}$$

Eighteen years before, LOTKA (1910) had already discussed this model from a reaction kinetic point of view with respect to its relations to autocatalytic processes. We shall come back to this latter interpretation later in this section and in Chapter 6. For an extensive study of the Volterra-Lotka model, the reader is referred to the monographs of GOEL, MAITRA and MONTROLL (1971) and MAY (1973).

The term k_1X ($k_1 > 0$) in (2.44) describes the reproduction of the prey species due to a constant supply from some external food resource. Without a predator species being present, the prey X would exponentially increase in population (Malthusian population growth). The k_3-term ($k_3 > 0$) represents the loss rate of the prey species due to "collisions" with the predators. The ansatz for this loss rate proportional to the product of population densities XY may be understood by comparing it with a collision process in reaction kinetics. Accordingly the k_4-term ($k_4 > 0$) represents the gain rate of the predator species, whereas the k_2-term ($k_2 > 0$) describes that the predator species would exponentially die out in the absence of preys.

First of all we notice that a scale transformation $X = k\tilde{X}/k_4$, $Y = k\tilde{Y}/k_3$ with arbitrary $k > 0$ leads to a system of differential equation with the same structure as in (2.44) but $k_3 = k_4 = k$. Let us assume that this transformation has already been performed in (2.44) such that we may set $k_3 = k_4 = k$ for the following.

Next we calculate the steady state $\overline{X}$, $\overline{Y}$ of the system. We immediately obtain

$$\overline{X}_1 = k_2/k, \quad \overline{Y}_1 = k_1/k \quad \text{or} \quad \overline{X}_2 = \overline{Y}_2 = 0 \ . \tag{2.45}$$

Our model shows "multi-stationarity": there are two steady states $\overline{X}_1$, $\overline{Y}_1$ and $\overline{X}_2$, $\overline{Y}_2$, among which the latter one, however, is "ecologically degenerate", since no population would be present at all. In order to decide whether the steady states are stable, we carry out a differential analysis in the vicinity of the steady states like that in Sections 2.2 and 2.3. Writing $\delta X = X - \overline{X}$, $\delta Y = Y - \overline{Y}$ where $\overline{X}$, $\overline{Y}$ may be either of the states in (2.45) and observing $\delta(XY) = \overline{X}\delta Y + \overline{Y}\delta X$ in the differential vicinity of $\overline{X}$, $\overline{Y}$ we obtain from (2.44)

$$\left.\begin{aligned}\frac{d}{dt}\,\delta X &= k_1\delta X - k(\overline{X}\delta Y + \overline{Y}\delta X) \\ \frac{d}{dt}\,\delta Y &= -k_2\delta Y + k(\overline{X}\delta Y + \overline{Y}\delta X)\end{aligned}\right\} . \tag{2.46}$$

Let us first investigate the ecologically degenerate state $\overline{X}_2 = \overline{Y}_2 = 0$:

$$\frac{d}{dt}\,\delta X = k_1 \delta X \quad , \qquad \frac{d}{dt}\,\delta Y = -\,k_2 \delta Y \quad . \tag{2.47}$$

The solutions of (2.47) are

$$\delta X(t) = \delta X(0)e^{k_1 t} \quad , \qquad \delta Y(t) = \delta Y(0)e^{-k_2 t} \quad . \tag{2.48}$$

The state $\overline{X}_2 = \overline{Y}_2 = 0$ shows "saddle-point behaviour": whereas it is stable in Y-direction, it is unstable in X-direction. An arbitrarily small initial fluctuation $\delta X(0) > 0$ leads to a further increase in X which means that $\overline{X}_2 = \overline{Y}_2 = 0$ as a whole is unstable.

Insertion of the steady state $\overline{X}_1$, $\overline{Y}_1$ into (2.46) gives

$$\frac{d}{dt}\,\delta X = -k_2 \delta Y \quad , \qquad \frac{d}{dt}\,\delta Y = k_1 \delta X \quad . \tag{2.49}$$

Eqs. (2.49) are the equations of motion of an harmonic oscillator with the general solution

$$\left.\begin{aligned} \delta X &= (\delta R/\sqrt{k_1}) \;\; \cos(\omega t + \varphi) \\ \delta Y &= (\delta R/\sqrt{k_2}) \;\; \sin(\omega t + \varphi) \end{aligned}\right\} \tag{2.50}$$

where $\omega = \sqrt{k_1 k_2}$ and δR and φ are constants of integration. The trajectories of the deviations and hence of the population variables are elliptic orbits in the X-Y-plane. The population variables X, Y rotate around the steady state in the manner of an harmonic oscillator with frequency ω. The axes of the elliptic orbit, $\delta R/k_1$ and $\delta R/k_2$, depend on the initial conditions.

The mathematical background of the periodic oscillations around the steady state $\overline{X}_1$, $\overline{Y}_1$ is a conservation law which plays quite an analogous role for the equations of motion (2.44) of the Volterra-Lotka model as the conservation of energy in the ordinary undamped harmonic oscillator in classical mechanics. The reader immediately verifies that the quantity

$$H = k(X + Y) - k_1 \ln Y - k_2 \ln X \tag{2.51}$$

is conserved if X and Y vary with time according to (2.44). This finding is exact and not restricted to differential vicinities around the steady state as (2.46). Thus (2.51) for $H = H_0 = \text{const}$ gives the general form of the trajectories of the system in the X-Y-plane. The reader may also easily verify that (2.51) for $H = H_0 = \text{const}$ describes closed trajectories which become elliptic orbits in the differential vicinity of the steady state $\overline{X}_1, \overline{Y}_1$.

The population dynamic processes which underly the Volterrra-Lotka model can also be given a reaction kinetic interpretation by writing

$$\left.\begin{array}{lr} A + X \rightleftharpoons 2X & (1) \\ X + Y \rightleftharpoons 2Y & (2) \\ Y \rightleftharpoons B & (3) \end{array}\right\} . \qquad (2.52)$$

Reaction (1) describes the reproduction of X maintained by some external resource A of food, reaction (2) describes the collisions between X and Y which lead to a reproduction of Y and reaction (3) describes the extinction of Y. Reactions (1) and (2) are autocatalytic: X and Y catalyze their own production. Let us assume that the reaction rates in (2.52) satisfy a product ansatz as introduced in (2.4):

$$\left.\begin{array}{l} J_1 = \gamma_1 AX - \gamma_1' X^2 \\ J_2 = \gamma_2 XY - \gamma_2' Y^2 \\ J_3 = \gamma_3 Y - \gamma_3' B \end{array}\right\} \qquad (2.53)$$

where the γ_i and γ_i' are the forward and reverse reaction rate constants, respectively. The complete differential equations for the time changes of the population densities now read

$$\frac{dX}{dt} = J_1 - J_2 \quad , \qquad \frac{dY}{dt} = J_2 - J_3 \quad . \qquad (2.54)$$

Comparing (2.54) with the original model equations (2.44) we see that both systems become identical if we set $k_1 = \gamma_1 A$, $k = k_3 = k_4 = \gamma_2$, $k_3 = \gamma_3$ and $\gamma_i' = o$ for $i = 1,2,3$. From a reaction kinetic point of view, $\gamma_i' = 0$ could be interpreted as an approximation for an extreme nonequilibrium situation such that the reverse fluxes are vanishingly small. Although reverse fluxes do not bear a direct population dynamic meaning, it is quite interesting to study the formal consequences if they are retained in (2.54). For example, the Malthusian growth rate term of X in the reaction (1) of (2.52) would then be replaced by

$$\frac{dX}{dt} = J_1 = \gamma_1 AX\left(1 - \frac{X}{X_s}\right) \qquad (2.55)$$

where $X_s = \gamma_1 A/\gamma_1'$. For small values of X, $X \ll X_s$, (2.55) describes the onset of an exponentially increasing population density of X with a rate $k_1 = \gamma_1 A$. As X approaches the value of X_s, the effective reproduction rate $\gamma_1 A(1 - X/X_s)$ tends to zero such that X saturates at $X = X_s$. This behaviour is easily confirmed by an exact analytic solution of (2.55).

Since saturation of a reproducing population density is what happens in real ecosystems rather than unlimited reproduction, (2.55) is often used as a simulation model, the first time more than 100 years ago by VERHULST (1845, 1847). The stabilization tendency of reverse rates in autocatalytic reactions and their indirect but meaningful ecological implications is quite a general phenomenon as we will see later in Chapter 6. Corresponding terms as generated by the reverse rates of reaction fluxes such as J_2 and J_3 in (2.53) are less familiar in population dynamics but lead to quite similar consequences. As LEISEIFER (1975) has shown, the inclusion of a reverse rate in one single flux of an extended Volterra-Lotka system with an arbitrary number of interacting species leads to an asymptotic stabilization of the steady state such that the originally undamped periodic oscillations without reverse rates become damped and tend to the steady state as $t \longrightarrow \infty$. This result distinguishes the conservative oscillations of the original Volterra-Lotka model principally from a type of periodic oscillations to be discussed in the next section.

2.7 A Model for the Control of Metabolic Reaction Chains

The model that we have discussed in Section 2.2 for enzyme-catalyzed reactions showed saturation of the reaction flux J at high values of the substrate concentration S. Clearly, this saturation phenomenon is due to the limited total amount of enzyme. It may also be considered as some effective control of the reaction flux which prevents an overstrain of the system with increasing substrate concentrations.

Quite a lot of reaction chains within the network of biochemical pathways show a much more efficient control mechanism in that some intermediate product S_n after n steps of the chain blocks or inhibits an enzyme E which catalyzes the first step of the chain:

$$\left.\begin{array}{c} S + E \rightleftharpoons S_1 + E \\ S_1 \rightleftharpoons S_2 \rightleftharpoons S_3 \rightleftharpoons \dots \rightleftharpoons S_n \rightleftharpoons P \\ E + \rho S_n \rightleftharpoons E^* \end{array}\right\} \quad . \tag{2.56}$$

In (2.56), S and P are the initial substrate and the final product of the chain, respectively, which are kept at fixed concentrations. In the inhibition reaction, a numer of ρ molecules of the intermediate product S_n is assumed to transfer the enzyme from its active configuration E to its inactived configuration E*.

The model in (2.56) has been investigated by GRIFFITH (1968), HUNDING (1974) and TYSON (1975). In contrast to the excitation models in Sections 2.4 and 2.5, it shows a negative feedback: a sudden increase of S leads to stepwise delayed increases of the chain products S_1, S_2, ... S_n until finally S_n reduces the enzyme E involved in the first step such that now the chain products S_1, S_2, ... S_n will decrease, etc. From this qualitative consideration we expect that the system will be able to perform oscillatory changes of its state. We shall convince ourselves by a mathematical analysis that this will really be possible; however, we will also learn that these oscillatory motions are basically different from those of the undamped Volterra-Lotka model.

In our analysis we follow the work of HUNDING (1974) and first make three approximations which are not essential for the qualitative behaviour of the model but make it mathematically tractable. The first approximation is the assumption that the inhibition reaction, i.e., the third line of (2.56), is so rapid compared with the chain reactions that it may be evaluated at its steady state, which in this particular case means vanishing inhibition reaction flux, such that

$$E^* = K(S_n)^\rho E \tag{2.57}$$

where K is the ratio of the rate constants in the inhibition reaction flux. Since in this case the steady state flux vanishes, (2.57) coincides with a thermodynamic equilibrium condition. The total amount of enzyme is conserved in (2.56), $E + E^* = E_0 = \text{const}$, which together with (2.57) yields

$$E = E_0 \cdot \left|1 + K(S_n)^\rho\right|^{-1} \quad . \tag{2.58}$$

In the second approximation we assume that the chain reactions are so far from equilibrium that only the forward rates leading from S to P have to be taken into account such that together with (2.58) the fluxes J_i from S_i to S_{i+1} are given as

$$\left.\begin{aligned} J_0 &= k_0 SE = k_0 E_0 S\left[1 + K(S_n)^\rho\right]^{-1} \\ J_i &= k_i S_i \qquad \text{for} \qquad i = 1,2, \ldots n \end{aligned}\right\} \quad . \tag{2.59}$$

Finally it is assumed that the constants k_1, k_2, ... k_n are equal: $k_1 = k_2 = \ldots = k_n \equiv k$. The time changes of S_i are now given as

$$\frac{dS_i}{dt} = J_{i-1} - J_i \qquad i = 1,2, \dots n \quad . \tag{2.60}$$

Let us first calculate the steady state S_i for which $dS_i/dt = 0$ and hence $\bar{J}_o = \bar{J}_1 = \dots = \bar{J}_n \equiv \bar{J}$ such that

$$k_oE_oS\left[1 + K(\bar{S}_n)^\rho\right]^{-1} = k\bar{S}_1 = \dots = k\bar{S}_n \quad . \tag{2.61}$$

Introducing a new parameter $Q = k_oE_oS/k\bar{S}_n$ we immediately obtain from (2.61)

$$Q(Q-1)^{1/\rho} = k_oE_oSK^{1/\rho}/k \tag{2.62}$$

and for the steady state flux

$$\bar{J} = k(Q-1)^{1/\rho}/K^{1/\rho} \quad . \tag{2.63}$$

In order to calculate $\bar{J}$ as a function of the substrate concentration S, we first have to solve (2.62) for Q and then to insert the result into (2.63). The reader may verify that for small values of S we have $\bar{J} \sim S$. For increasing S, this linear relation is flattened, however, not saturated as in the model of Section 2.2. Nevertheless, the tendency of the control mechanism evidently is the same.

Let us now turn to the time-dependent properties of the model. Again we restrict ourselves to differential fluctuations $\delta S_i = S_i - \bar{S}_i$ around the steady state. From (2.60) we obtain

$$\frac{d}{dt}\,\delta S_i = \delta J_{i-1} - \delta J_i \qquad i = 1,2, \dots n \tag{2.64}$$

and from (2.59)

$$\delta J_o = -\alpha k \delta S_n \; ; \qquad \delta J_i = k\delta S_i \; , \qquad i = 1,2, \dots n \tag{2.65}$$

where

$$\alpha = \frac{\rho K k_oE_oS(\bar{S}_n)^{\rho-1}}{k\left[1 + K(\bar{S}_n)^\rho\right]^2} = \rho\,\frac{Q-1}{Q} \quad . \tag{2.66}$$

The calculation of δJ_o involves a differentiation with respect to S_n and the replacement of S_n by its steady state value $\bar{S}_n$ expressed by Q. Combining (2.64) and (2.65) leads to the system of equations of motion

$$\left.\begin{aligned} \frac{d}{dt}\,\delta S_1 &= -\alpha k\delta S_n - k\delta S_1 \\ \frac{d}{dt}\,\delta S_i &= k\delta S_{i-1} - k\delta S_i \qquad i = 2,3,\ \dots\ n \end{aligned}\right\} . \tag{2.67}$$

Introducing a new variable ξ_i by

$$\xi_i = e^{kt}\delta S_i\ , \qquad i = 1,2,\ \dots\ n \tag{2.68}$$

we reduce (2.67) to

$$\frac{d\xi_1}{dt} = -\alpha k\xi_n\ , \qquad \frac{d\xi_i}{dt} = k\xi_{i-1} \qquad i = 2,\ \dots\ n\ . \tag{2.69}$$

From (2.69) we obtain for the nth order time derivative of ξ_n

$$\left(\frac{d}{dt}\right)^n \xi_n = -\alpha k^n \xi_n\ . \tag{2.70}$$

Eq. (2.70) is immediately solved by

$$\xi_n(t) = e^{\lambda t}\xi_n(0)\ , \qquad \lambda^n = -\alpha k^n \tag{2.71}$$

such that from (2.68)

$$\delta S_n(t) = e^{(\lambda - k)t}\delta S_n(0) \tag{2.72}$$

λ may be any of the roots of the nth order equation in (2.71)

$$\lambda = \lambda_m = k|\alpha|^{1/n}\exp\left(i\,\frac{\pi}{n} + i\,\frac{2\pi}{n}\cdot m\right), \qquad m = 0,1,\ \dots\ n-1 \tag{2.73}$$

which lie on a circle in the complex λ-plane with a center at $\lambda = 0$ and a radius $r = k|\alpha|^{1/n}$. The general solution of $\delta S_n(t)$ is a linear combination of the expressions (2.72) for the various values of $\lambda = \lambda_m$ as given in (2.73). Whether this solution is stable or not depends on the sign of the maximum value of the real part of $\lambda_m - k$ which is obtained for m = 0 or m = n - 1

$$\text{Max Re}\left\{(\lambda_m - k)\right\} = k\left(|\alpha|^{1/n}\cos\frac{\pi}{n} - 1\right)\ . \tag{2.74}$$

Inserting α from (2.66) we thus have stability if

$$\left|\rho \frac{Q-1}{Q}\right|^{1/n} \cos\frac{\pi}{n} < 1 \tag{2.75}$$

and instability if

$$\left|\rho \frac{Q-1}{Q}\right|^{1/n} \cos\frac{\pi}{n} > 1 \quad . \tag{2.76}$$

A necessary condition for the inequality (2.76) to be realizable is $\rho^{1/n}\cos(\pi/n) > 1$ which means, e.g., $\rho > 8$ for $n = 3$ and $\rho > 4$ for $n = 4$. If this condition is fulfilled, a sufficiently large value of Q, i.e., a sufficiently large substrate concentration S, cf. (2.62), will then cause at least two of the modes of (2.73) to leave the steady state in a spiral motion. The question of what eventually happens to these instable modes cannot be answered within the framework of our differential analysis of the steady state. On the other hand, we do not expect that these modes really diverge to infinite values of S_n since the feedback coupling of our model will prevent such an explosion. As a matter of fact, one can show, (cf. TYSON (1975)), that the instable modes approach a so-called limit cycle, i.e., a stable, undamped periodic oscillation around the steady state. The instable spiral motion, which we have found in our differential analysis, thus represents the initial phase of the evolution towards the limit cycle.

Let us conclude our considerations by pointing out what distinguishes a limit cycle principally from the periodic oscillation in the Volterra-Lotka model without saturation terms. In the latter case, the periodic oscillation is the consequence of a conservation law. This means that it appears under any external conditions and that its trajectory depends on the initial condition. A limit cycle, however, is a unique asymptotic trajectory which is independent of the initial condition of a particular motion and into which all motions tend with $t \longrightarrow \infty$ irrespective of their individual history. Moreover, a limit cycle only appears under certain minimal external conditions as we have seen in context with the inequality (2.76). We shall come back to the phenomenon of limit cycles in a more general context in Section 6.4.

Problems

1) If the reaction from substrate S to the product P is assumed to occur while S or P are bound to the enzyme E, the reaction scheme of (2.1) should be extended to

$$E + S \rightleftharpoons [ES] \rightleftharpoons [EP] \rightleftharpoons E + P$$

Develop a model theory for the extended scheme in analogy to Section 2.2. Are the results of the extended theory essentially different from or qualitatively the same as that of Section 2.2?

2) Design a network for two different membranes in series in analogy to Fig. 3. For which situation could this arrangement serve as a model? What are the fundamental steady state and relaxation properties?

3) Simplify the Hodgkin-Huxley model in such a way that it contains only one single ionic channel which is controlled by one charged particle in a unimolecular reaction. Assume a voltage dependence of the rate constants in the form $k = k_0 \exp(\alpha V)$, $k' = k_0' \exp(-\alpha V)$. Discuss the possible steady state I-V-characteristics of this reduced model. Extend this discussion to the original Hodgkin-Huxley model.

4) Formulate the full equation of motion for the Ising excitation model of Section 2.5 by inserting (2.42) into (2.38). Perform a differential analysis in the vicinity of the various steady states and prove that the steady state on the intermediate branch in Fig. 6 is instable.

5) Perform the transformation

$$X = e^u , \qquad Y = e^v$$

in (2.44) and show that (2.44) can then be generated from H of (2.51) as a Hamilton function such that

$$\frac{du}{dt} = - \frac{\partial H}{\partial v} , \qquad \frac{dv}{dt} = \frac{\partial H}{\partial u}$$

6) Introduce an intermediate substrate-enzyme complex [SE] into the first step of the reaction chain of Section 2.7 similarly as in the model of Section 2.2. Prove that now for increasing values of the substrate concentration S the steady state flux $\bar{J}$ saturates. Calculate the saturation value of $\bar{J}$.

3. Thermodynamics

3.1 Thermodynamic Systems

Thermodynamics is concerned with systems with very large numbers of microscopic degrees of freedom. Consider for example one mole of a monatomic gas enclosed in a fixed volume. The number of atoms in this system is given by Advogadro's number $L = 6{,}025.10^{23}$. The mechanical state of motion of each of the atoms is characterized by each 3 coordinates for its position and its velocity. Hence, the number of the so-called microscopic degrees of freedom of the total system is $f = 6\ L$.

Also, a biological cell has to be considered as a thermodynamic system. Its number f of microscopic degrees of freedom is still larger than that of the above example even if the same number L of particles were present in the cell since the whole variety of chemical reactions including all metabolic and genetic phenomena generates additional degrees of freedom.

The number of microscopic states, the so-called micro-states, which are accessible to thermodynamic systems is even more impressive. To give a very simple example let us consider a system of L particles each of which can exist in only two different micro-states as in the case of the spin variable of an electron. Since the particles can take on their two states independent of each other, the number of micro-states of the total system is given by

$$2^L \approx 10^{2 \cdot 10^{23}} \quad .$$

In contrast to this microscopic description of many-particle systems, thermodynamics aims at a macroscopic description of such systems. Typical macroscopic variables are for example the volume of the system, its total mass, the number of moles of its chemical components and its total energy. In any case, the number of thermodynamic or macroscopic variables is much less than the number of the microscopic degrees of freedem. Hence, the transition from a microscopic to a macroscopic description involves a drastic reduction of the information about the system. This means that any particular macro-state must be realized by a very large number of micro-states among which the system rapidly fluctuates. In the course of a macroscopic measurement of the system, however, such microscopic fluctuations will not be observed.

The precise formulation of the relationship between the microscopic and macroscopic description of many-particle systems is the subject of statistical physics. Actually, this program is practicable only for rather simple and nearly homogeneous systems and only for macro-states very near to thermodynamic equilibrium. The examples of biological models, however, which we have discussed in Chapter 2, have shown us that for a physical analysis of biological systems we are particularly interested in complex, inhomogeneous systems very far from thermodynamic equilibrium. Thus we can not expect much progress from an analysis which starts at the very microscopic level of description. Instead, we shall attempt to give a brief formulation of the basic laws of thermodynamics and irreversible thermodynamics as far as needed for our purposes. In doing so, we shall incidentally recall the relation between thermodynamics and its microscopic basis. This will be the program of the next sections of this chapter.

As far as equilibrium thermodynamics is regarded, we shall closely follow the lines of CALLEN's (1960) book which is particularly recommended for further reading[1]. An application of thermodynamics especially to chemical processes is presented in detail in the book of PRIGOGINE and DEFAY (1954). For an introduction into thermodynamics of irreversible processes the interested reader is referred to the books of de GROOT (1951), de GROOT and MAZUR (1962) and KATCHALSKY and CURRAN (1967).

3.2 The First Law of Thermodynamics

The first law of thermodynamics expresses the energy-conservation principle. There are various ways upon which a thermodynamic system can energetically interact with its surroundings and thereby exchange energy. Consider a gas enclosed in a cylinder the interior of which is on one side separated from the exterior by a movable piston. This system exchanges mechanical work with the exterior if the piston is pushed into or pulled out of the cylinder. If the piston or the walls of the cylinder are perforated, the system will exchange matter with its surroundings, and, coupled to this flux of matter, again energy will be exchanged.

Energy exchange coupled to the flux of matter represents the most important energy transfer process in biological systems. Let us consider a biological cell as a thermodynamic system. The cell membrane which separates the cell from its surrounding is specifically permeable to certain ions and molecules. Every flux of such ions or molecules across the membrane is accompanied by an exchange of energy between the system and its surroundings.

---

[1]An introduction into thermodynamics from a particular biological point of view is given in two books by HILL (1968) and MOROWITZ (1970)

There is still a further type of energy exchange which we have not yet mentioned but which play a dominant role in the following, namely, that of heat transfer. At this point, we can give only a preliminary understanding of the thermodynamic phenomenon of heat, namely, by saying that heat transfer is the macroscopic expression of energy exchange between the microscopic degrees of freedom of the system and its surroundings. It is evident that the microscopic degrees of freedom do carry energy for example in the form of kinetic energy of the molecules and of potential energy due to the interaction between the molecules. It is also quite natural to assume that this kind of energy can be directly exchanged between the system and its surroundings.

The first law of thermodynamics now states that there exists a macroscopic thermodynamic variable U, called the internal energy, such that the net amount of energy exchanges of the system with its surroundings is stored into or subtracted from the internal energy. Let us consider an arbitrary transition from some macro-state A to some macro-state B during which energy is exchanged with the surroundings in some arbitrary way. The first law can then be expressed as

$$U(B) - U(A) = \sum_i \Delta W_i(A \longrightarrow B) \tag{3.1}$$

where U(A) and U(B) are the values of the internal energies in the macro-states A and B, respectively, and $\Delta W_i(A \longrightarrow B)$ is the amount of energy exchange of type i during the process $A \longrightarrow B$. The single contributions $\Delta W_i(A \longrightarrow B)$ may have positive or negative values depending on the detailed circumstances of the process $A \longrightarrow B$.

The important point of (3.1) is the fact that it not merely defines a quantity U but actually represents an autonomous statement by saying that the left-hand side is independent of the history of the particular process $A \longrightarrow B$ but depends only on the initial and the terminal macro-states A and B of the process. Such variables which are uniquely related to macro-states and independent of the processes between the macro-states are called state variables. Thus, the first law of thermodynamics can also be expressed by the statement that the internal energy U is a state variable.

On the other hand, the single contributions $W_i(A \longrightarrow B)$ to the right-hand side of (3.1) are not state variables. For example, the mechanical work which is needed to compress a gas enclosed in a cylinder depends on the compression velocity and also on the physical nature of the walls of the cylinder, namely, whether the walls allow a transfer of heat during the compression or are thermally isolating. It is only the sum of all contributions on the right-hand side of (3.1) which is independent of the history and the circumstances of the process.

Another way of formulating the first law is to replace the finite processes between macro-states A and B in (3.1) by infinitesimal or differential processes such that

$$dU = \sum_i dW_i \quad . \tag{3.2}$$

According to our above discussion we now have to bear in mind that only the left-hand side of (3.2) is a total differential of a state variable U, but not necessarily the single differential expressions dW_i on the right-hand side.

If the system can exchange energy only in form of mechanical work dA and heat dQ as in the case of the gas enclosed in a nonperforated cylinder, (3.2) reads

$$dU = dA + dQ \quad . \tag{3.3}$$

For slow velocities of compression or dilation of the gas, so-called quasi-static processes, dA is given as

$$dA = - pdV \tag{3.4}$$

where p is the pressure, dV is the change of volume and the sign ensures that $dA > 0$ for a compression $dV < 0$. The first law now reads

$$dU = - pdV + dQ \quad . \tag{3.5}$$

For a system which can exchange energy only as heat or coupled to material fluxes, (3.2) is written as

$$dU = dQ + \sum_i \mu_i dN_i \tag{3.6}$$

where dN_i denotes the infinitesimal change of the number of moles of the component i and μ_i is the so-called chemical potential of component i, i.e., the amount of energy exchange associated with the exchange of one mole of the component i. Eq. (3.6) is the appropriate form of the first law if applied to biological systems where the most frequent forms of energy exchange are those of heat and coupled to material exchange.

3.3 The Second Law of Thermodynamics

Let us consider a thermodynamic system in arbitrary contact with its surroundings and let A_0 be its macro-state at some time t_0 which due to the contact with the surroundings may include spatial inhomogeneties of the components of the system as well as macroscopic turbulent flows or differences of the pressure within the system, etc. Now, at time t_0 the system is instantaneously isolated from its surrounding such that no further energy exchange can take place and the value of the internal energy U remains constant. What will be observed then for times t later than t_0 is a spontaneous evolution of the system towards a macro-state in which the components are homogeneously distributed, the turbulent flows are damped out and the pressure differences are compensated, etc. We call this experimentally observed terminal macro-state the "thermodynamic equilibrium".

In order to characterize an equilibrium state let us simplify the situation by restricting ourselves to systems consisting of arbitrary chemical components i = 1,2, ... with mole numbers N_1, N_2 ... enclosed in a volume V. We now generalize the experimental finding of terminal macro-states in isolated systems by stating that there exist particular macro-states of thermodynamic systems, the so-called equilibrium states, which are uniquely determined by the values of the macro-variables internal energy U, volume V and mole numbers N_1, N_2 ...

The macro-variables U, V, N_1, N_2, ... have an important physical property in common. If we form a new thermodynamic system by combining two identical systems, the new values of each of the variables U, V, N_1, N_2, ... will be doubled as compared to that of the single system. Macro-variables with this property are called extensive variables. Macro-variables with values that are not affected by changes of the size of the system are called intensive variables. As examples for intensive variables consider quotients of extensive variables like $c_i = N_i/V$, the concentration of the component i (moles per liter) or U/V, the internal energy per unit volume.

After having postulated the existence of equilibrium states, we shall try to formulate a criterion which distinguishes an equilibrium state from nonequilibrium macro-states. Recalling the intuitive description of the evolution towards the equilibrium state at the beginning of this section, one could be inclined to characterize equilibrium as the most probable state of the system that is compatible with the given values of U, V, N_1, N_2, ... Indeed, let us compare the equilibrium state E with some nonequilibrium state A belonging to the same values U, V, N_1, N_2, ... Let A be defined as a state where all particles of the system are contained in a partial volume V_1 of the total volume V. The probability of a single particle of the system with volume V to stay in V_1 is V_1/V such that for independent particles the probability of the state A compared to the equilibrium E with its homogeneous distribution of particles is given as

$$p = (V_1/V)^M \tag{3.7}$$

where M is the total number of particles. Inserting, for example, $M = L \approx 6 \cdot 10^{23}$ and excluding only 1 % of the total volume V for the particles in the state A, i.e., $V_1/V = 0.99$, gives

$$p = (0.99)^{6 \cdot 10^{23}} \approx 10^{-2.6 \cdot 10^{21}} \quad .$$

We now understand that it is the larger number of particles or, in other words, the large number of microscopic degrees of freedom in thermodynamic systems which makes even very slight deviations from the equilibrium state extremely improbable. The spontaneous evolution of an isolated system towards its equilibrium state is nothing else than the transition to the most probable one. The equilibrium state is distinguished from the nonequilibrium states by the condition of maximum probability and due to the large number of microscopic degrees of freedom this maximum is extremely sharp.

In phenomenological thermodynamics this maximum principle is expressed in terms of a quantity called the entropy S. The entropy S is related to the probability p of the statistical treatment by

$$S = k \ln p \tag{3.8}$$

the constant k depending on the choice of units, see below. Since the logarithm ln is a monotonically increasing function of its argument, it is evident that S is maximal if p is maximal and vice versa.

We are now in a position to formulate the second law of thermodynamics in the following way: there exists a thermodynamic macro-variable entropy S such that a) S is an extensive variable, b) S is maximal in the equilibrium state defined by $U, V, N_1, N_2, \ldots$ in isolated systems and c) for equilibrium states which differ by their values for U but have the same values of $V, N_1, N_2 \ldots$, the entropy S is a monotonically increasing function of U.

Property a) follows directly from (3.8): if two systems are combined, the corresponding probabilities have to be multiplied such that S of the combined system becomes the sum of the entropies of the partial systems.

Property b) implies that the maximum value of S for an equilibrium state defined by given values for $U, V, N_1, N_2 \ldots$ may be considered as a unique function of $U, V, N_1, N_2, \ldots$.

$$S = S(U, V, N_1, N_2, \ldots) \quad . \tag{3.9}$$

Property c) now expresses that

$$\left(\frac{\partial S}{\partial U}\right)_{V,N_i} > 0 \tag{3.10}$$

where the subscripts V, N_i indicate that the variables V, N_1, N_2, ... are to kept fixed when performing the differentiation with respect to U. Property c) or (3.10) may again be understood on the basis of the statistical interpretation of the entropy. The larger the value of the total internal energy U of the system, the larger will be the number of micro-states that are accessible to the system. As a consequence, the number of micro-states by which an equilibrium state is realized will increase with the value of U. Evidently, this implies that the larger the value of U, the larger will be the maximum value of the relative probability for the equilibrium and thus the value of the equilibrium entropy S.

3.4 Temperature and Transfer of Heat

The aim of this section is to justify the quite formal definition of the (absolute) temperature T by

$$\frac{1}{T} = \left(\frac{\partial S}{\partial U}\right)_{V,N_i} \quad , \tag{3.11}$$

to investigate conditions under which a system is in equilibrium with respect to transfer of heat, and to inquire what happens if such conditions are not satisfied. Let us recall that S in (3.11) is understood as the maximum value of the entropy in an equilibrium state as formulated in context with (3.9) and that V and N_i for i = 1, 2, ... are to be kept constant when performing the differentiation with respect to U. From the definition (3.11) it is evident that temperature is an intensive variable since it is defined as a quotient of two extensive variables.

In order to present the essential point of our considerations as clearly as possible, let us start with a very simple situation in which two systems (1) and (2), each of which is in internal equilibrium, are brought into thermal contact. The combined system (1) + (2) as a whole is assumed to be isolated from any further energetic interaction with the surroundings. By thermal contact we mean that there may be energetic exchange between systems (1) and (2) only by transfer of heat, but not by transfer of any other form of energy. Without further specifications, systems (1) and (2) apparently need not be in equilibrium with respect to each other or, in other words, the combined system as a whole need not be in an equilibrium state. Our first problem is to find a condition for the equilibrium of the combined system by making use of the first and second law of thermodynamics.

The second law tells us that subsystems (1) and (2) are each characterized by their equilibrium entropies $S^{(1)}$ and $S^{(2)}$ as functions of their internal energies $U^{(1)}$ and $U^{(2)}$:

$$S^{(1)} = S^{(1)}\left(U^{(1)}\right) \quad , \quad S^{(2)} = S^{(2)}\left(U^{(2)}\right) \quad . \tag{3.12}$$

We have dropped the variables V, N_1, N_2, ... since their values are kept constant for each of the subsystems (1) and (2). As entropy is an extensive variable, the entropy S of the combined system is given as

$$S = S^{(1)}\left(U^{(1)}\right) + S^{(2)}\left(U^{(2)}\right) \quad . \tag{3.13}$$

As the internal energy is likewise extensive, the total internal energy U of the combined system is given as the sum of $U^{(1)}$ and $U^{(2)}$. According to the assumption that the combined system as a whole is isolated, we derive from the first law of thermodynamics

$$U = U^{(1)} + U^{(2)} = \text{const} \quad . \tag{3.14}$$

Let $\delta U^{(1)}$ and $\delta U^{(2)}$ be infinitesimal changes, so-called variations of $U^{(1)}$ and $U^{(2)}$, respectively. Due to (3.14) we have

$$\delta U^{(1)} + \delta U^{(2)} = 0 \quad . \tag{3.15}$$

Clearly, $\delta U^{(1)}$ is the amount of heat transferred from system (1) to system (2), and vice versa $\delta U^{(2)}$.

The infinitesimal variation of the total entropy S due to a variation of $U^{(1)}$ and $U^{(2)}$ is calculated as

$$\delta S = \frac{\partial S^{(1)}}{\partial U^{(1)}} \delta U^{(1)} + \frac{\partial S^{(2)}}{\partial U^{(2)}} \delta U^{(2)} = \left(\frac{1}{T^{(1)}} - \frac{1}{T^{(2)}}\right) \delta U^{(1)} \tag{3.16}$$

where use has been made of the definition of temperature, (3.11), and of the condition (3.15). The necessary condition for the combined system to be in an equilibrium state is that S is maximal or, as a consequence, that $\delta S = 0$ from which we conclude

$$T^{(1)} = T^{(2)} \quad . \tag{3.17}$$

Thus, the quite formally introduced temperature T has the property that subsystems (1) and (2) are in thermal equilibrium with respect to each other if their temperatures are equal. This is a first justification of our definition (3.11).

If condition (3.17) is not fulfilled, i.e., if $T^{(1)} \neq T^{(2)}$, the spontaneous variations $\delta U^{(1)}$ and $\delta U^{(2)}$ will be such that $\delta S > 0$, since we observe that isolated systems spontaneously evolve towards an equilibrium state and the entropy is maximal in the equilibrium state:

$$\delta S = \left(\frac{1}{T(1)} - \frac{1}{T(2)} \right) \delta U^{(1)} > 0 \qquad \text{if} \qquad T^{(1)} \neq T^{(2)} \quad . \tag{3.18}$$

Interpreting the spontaneous variations as continuous processes in time, we may also write

$$\dot{S} = \Delta \frac{1}{T} \cdot I_U > 0 \qquad \text{if} \qquad \Delta \frac{1}{T} \neq 0 \tag{3.19}$$

where $\dot{S}$ is the change of entropy per time, the so-called entropy production, I_U is the amount of internal energy per time transferred from (2) to (1), the so-called flux of internal energy, and

$$\Delta \frac{1}{T} = \frac{1}{T^{(1)}} - \frac{1}{T(2)} \quad . \tag{3.20}$$

Eq. (3.19) tells us that $I_U > 0$ if $T^{(1)} < T^{(2)}$ and vice versa, which means that heat is always transferred from higher to lower temperatures. This is a second justification of our definition (3.11).

The inequalities (3.18) or (3.19) determine a unique and irreversible direction of spontaneous evolutions in systems which are not in equilibrium. We see that such irreversible evolutions are always associated with a production of entropy. This entropy production $\dot{S}$ vanishes only if the temperatures of the two subsystems are equal. In this case, no spontaneous heat transfer will be observed such that $\Delta\,(1/T) = 0$ implies $I_U = 0$ and vice versa.

On the other hand, we might think of a so-called reversible transfer of heat as an idealized transfer for vanishingly small temperature differences. As a consequence, then also the flux of transferred internal energy will be vanishingly small such that the idealized reversible transfer is an infinitely slow process. Since the entropy production is given as the product of the difference of reciprocal temperatures and the heat flux, it is small of second order and may be neglected for idealized reversible processes in comparison with that of other processes in the system occurring at finite rates. As an example let us assume that in our above construction systems (1) and (2) each receive some further heat flux $I_U^{(1)rev}$ and $I_U^{(2)rev}$ in an idealized reversible way besides the mutual irreversible exchange such that

$$\left.\begin{aligned} \dot{U}^{(1)} &= I_U^{(1)rev} + I_U \\ \dot{U}^{(2)} &= I_U^{(2)rev} - I_U \end{aligned}\right\} \qquad (3.21)$$

and

$$\dot{S} = \frac{\dot{U}^{(1)}}{T(1)} + \frac{\dot{U}^{(2)}}{T(2)} = \frac{I_U^{(1)rev}}{T(1)} + \frac{I_U^{(2)rev}}{T(2)} + \Delta \frac{1}{T} \cdot I_U \quad . \qquad (3.22)$$

Eq. (3.22) shows us that there are now two contributions to $\dot{S}$, namely a reversible change of entropy

$$\dot{S}^{rev} = \frac{I_U^{(1)rev}}{T(1)} + \frac{I_U^{(2)rev}}{T(2)} \qquad (3.23)$$

which may have any sign depending on the reversible fluxes, and an irreversible change

$$\dot{S}^{irr} = \Delta \frac{1}{T} \cdot J_U > 0 \qquad (3.24)$$

which has the same form as previously and is always positive.

A further generalization of our considerations is concerned with the introduction of the concept of heat baths or thermo-states. If the temperature of system (2) is kept fixed by some appropriate external manipulation, perhaps by making system (2) infinitely large, system (2) is called a heat bath or a thermo-state for system (1). As before, system (1) is in equilibrium with its heat bath if its temperature equals that of the heat bath. This equilibrium condition can now equivalently be expressed as an extremum principle in terms of the macro-variables of only system (1). For simplicity of writing let us denote the variables of system (1) without superscripts in the following. The macro-variable which becomes extremal in the equilibrium is the so-called free energy of system (1) defined as

$$F = U - T_o S \qquad (3.25)$$

where T_0 is the temperature of the heat bath. As before, the macro-variable which can be exchanged between the system and its heat bath is U such that

$$\delta F = \delta U - T_0 \delta S = \left(1 - T_0 \frac{\partial S}{\partial U}\right)\delta U = - T_0\left(\frac{1}{T} - \frac{1}{T_0}\right)\delta U = - T_0 \delta S_{tot} \tag{3.26}$$

where we have inserted expression (3.18) for the total variation of entropy δS_{tot} of the system and its heat bath (T replaces $T^{(1)}$ and T_0 replaces $T^{(2)}$). In the equilibrium state, we have $\delta S_{tot} = 0$ and thus $\delta F = 0$. For spontaneous variations from nonequilibrium states towards the equilibrium, δS_{tot} is always positive and hence $\delta F < 0$ such that the free energy F of a system in thermal contact with a heat bath has a minimum in the equilibrium state.

The next step in our brief development of thermodynamics is to define the pressure p by

$$p = \left(T \frac{\partial S}{\partial U}\right)_{U,N_i} \quad . \tag{3.27}$$

By applying the foregoing considerations for the case of energy exchange to the situation where now energy and volume can be exchanged between two systems such that $U^{(1)} + U^{(2)} = U = \text{const}$ and $V^{(1)} + V^{(2)} = V = \text{const}$, the reader may readily prove that equilibrium is established now if

$$T^{(1)} = T^{(2)} \quad \text{and} \quad p^{(1)} = p^{(2)} \quad . \tag{3.28}$$

If condition (3.28) is violated, we again have spontaneous processes with an entropy production

$$\dot{S} = \Delta \frac{1}{T} \cdot I_U + \Delta \frac{p}{T} \cdot I_V > 0 \tag{3.29}$$

where $\Delta(p/T) = p^{(1)}/T^{(1)} - p^{(2)}/T^{(2)}$ and I_V is the volume flux from (2) to (1), i.e., the gain of volume of (1) at the expense of (2) or the loss of volume of (1) in favour of (2). System (2) now may act as a simultaneous heat and mechanical bath system for system (1) if its temperature and pressure are kept fixed. The equilibrium between system (1) and this kind of extended bath system can then be expressed by an extremum principle for the so-called free enthalpy or Gibb's' free energy

$$G = F + p_0 V = U - T_0 S + p_0 V \tag{3.30}$$

where T_o and p_o are the fixed values of temperature and pressure of the bath system and U, S and V are the internal energy, entropy and volume of system (1).

3.5 Chemical Potential and the Transfer of Matter

The most frequent thermodynamic situation of a biological system is that of an exchange of heat and matter whereas its volume is kept fixed. Let us therefore extend the considerations of the preceding sections particularly to this kind of exchange. We start by defining the chemical potential μ_i of the component i by

$$\mu_i = -T\left(\frac{\partial S}{\partial N_i}\right)_{U,V,N_j} \tag{3.31}$$

the subscripts indicating that U, V and all N_j with $j \neq i$ are kept fixed when differentiating with respect to N_i. We shall see that definition (3.31) agrees with our earlier definition in (3.6)

Let us again consider two subsystems (1) and (2) which can mutually exchange energy U and chemical components N_i but are isolated as a whole. Writing the entropies $S^{(1)}$, $S^{(2)}$ of systems (1), (2) as functions of $U^{(1)}$, $N_i^{(1)}$ and $U^{(2)}$, $N_i^{(2)}$, respectively, and taking the variation of the total entropy $S = S^{(1)} + S^{(2)}$ we obtain in analog with (3.16)

$$\delta S = \left(\frac{1}{T^{(1)}} - \frac{1}{T^{(2)}}\right)\delta U^{(1)} - \sum_i \left(\frac{\mu_i^{(1)}}{T^{(1)}} - \frac{\mu_i^{(2)}}{T^{(2)}}\right)\delta N_i^{(1)} \geq 0 \tag{3.32}$$

where use has been made of (3.31) and of the isolation conditions $\delta U^{(1)} + \delta U^{(2)} = 0$, $\delta N_i^{(1)} + \delta N_i^{(2)} = 0$ for all i = 1, 2, ... The volumes of systems (1) and (2) are kept fixed separately: $\delta V^{(1)} = \delta V^{(2)} = 0$. For continuous spontaneous processes towards the equilibrium we obtain from (3.32)

$$\dot{S} = \Delta \frac{1}{T} \cdot I_U - \sum_i \Delta \frac{\mu_i}{T} \cdot I_i \geq 0 \tag{3.33}$$

where I_U and I_i are the fluxes of energy and of the chemical component i from (2) to (1). From (3.32) or (3.33) we conclude the equilibrium conditions as

$$T^{(1)} = T^{(2)} \quad \text{and} \quad \mu_i^{(1)} = \mu_i^{(2)} \quad , \quad i = 1, 2, \ldots . \tag{3.34}$$

If just one pair of chemical potentials, say for i = 1, violates (3.34) whereas $\mu_i^{(1)} = \mu_i^{(2)}$ for $i \neq 1$ and $T^{(1)} = T^{(2)} \equiv T$ we have from (3.33)

$$\dot{S} = -\frac{1}{T}\left(\mu_1^{(1)} - \mu_1^{(2)}\right) \cdot I_1 \geq 0 \tag{3.35}$$

which tells us that there will be a flux of component i = 1 from the subsystem with the larger value of μ_1 to that with the lower value of μ_1. If, however, the equilibrium condition is simultaneously violated for two components, say for i = 1 and i = 2, but $\mu_i^{(1)} = \mu_i^{(2)}$ for $i \neq 1, 2$ and $T^{(1)} = T^{(2)} \equiv T$, we have

$$\dot{S} = -\frac{1}{T}\left(\mu_1^{(1)} - \mu_1^{(2)}\right) \cdot I_1 - \frac{1}{T}\left(\mu_2^{(1)} - \mu_2^{(2)}\right) \cdot I_2 \geq 0 \quad . \tag{3.36}$$

Now, we can no longer conclude that each of the fluxes of components 1 and 2 follows the direction from higher to lower values of its chemical potential. Due to some kind of an unknown internal coupling it might happen now that a single component flows uphill from a lower towards a higher value of its chemical potential, at least for a limited time interval until in the equilibrium state all fluxes will eventually come to rest.

A particularly important generalization in view of subsequent applications of thermodynamics is the inclusion of a transfer of electric charges between the subsystems. Assume that the components i = 1, 2, ... could also be ions and let z_i be the valency of the component i such that $z_i = +1, +2, \ldots$ denotes cations and $z_i = -1, -2, \ldots$ anions, whereas $z_i = 0$ stands for neutral molecules. If subsystems (1) and (2) now have different electric potentials $\Phi^{(1)}$ and $\Phi^{(2)}$, an electric field will be present in the boundary region between the subsystems and consequently an exchange of charged particles between the subsystems will be associated with a transfer of energy from the electric field into the total internal energy or vice versa. Instead of $\delta U^{(1)} + \delta U^{(2)} = 0$ we then have

$$\delta U = \delta U^{(1)} + \delta U^{(2)} = \sum_i z_i F\left(\Phi^{(2)} - \Phi^{(1)}\right) \cdot \delta N_i^{(1)} \tag{3.37}$$

where F is Faraday's number, i.e., the electric charge per mole of monovalent ions. Let us assume that the temperatures of the subsystems are equal. $T^{(1)} = T^{(2)} \equiv T$ Insertion of (3.37) into the variation of the total entropy then leads to

$$\delta S = -\frac{1}{T}\sum_i\left(\eta_i^{(1)} - \eta_i^{(2)}\right)\delta N_i^{(1)} \tag{3.38}$$

where

$$\eta_i^{(\alpha)} = \mu_i^{(\alpha)} + z_i F \Phi^{(\alpha)} \quad , \qquad \alpha = 1, 2 \tag{3.39}$$

is the so-called electrochemical potential of component i in subsystem (α). The subsystems are now in equilibrium with respect to an exchange of ions if their electrochemical potentials are equal. Although in general the presence of electric fields could also affect the values of the $\mu_i^{(\alpha)}$, it is usually assumed as an approximation that the $\mu_i^{(\alpha)}$ are independent of the electric potentials.

We conclude this section by asking for an extremum principle for the equilibrium of a system which is in material and heat contact with a bath system. Following the recipes of the preceding sections we easily prove that

$$\Psi = U - T_0 S - \sum_i \mu_{io} N_i \tag{3.40}$$

is minimal in the equilibrium state, where T_0 and μ_{io} are the temperature and the chemical potentials in the bath system.

3.6 Chemical Reactions

In addition to an external variation of the mole numbers by a transfer of molecules across the boundary of the system, there may also be an internal variation due to chemical reactions. Quite generally, a chemical reaction can be formulated as

$$\nu_{1\rho}^{f} X_1 + \nu_{2\rho}^{f} X_2 + \ldots \rightleftharpoons \nu_{1\rho}^{r} X_1 + \nu_{2\rho}^{r} X_2 + \ldots \tag{3.41}$$

where the X_i are the symbols of the components and $\nu_{i\rho}^{f}$ and $\nu_{i\rho}^{r}$ are the forward and reverse stoichiometric coefficients of component i in the reaction numbered by ρ. For example, in the reaction

$$N_2 + 3H_2 \rightleftharpoons 2NH_3 \tag{3.42}$$

with $X_1 = N_2$, $X_2 = H_2$, $X_3 = NH_3$ the forward and reverse stoichiometric coefficients are (1, 3, 0) and (0, 0, 2), respectively. An example for a reaction with simultaneously nonvanishing forward and reverse coefficients is

$$S + E \rightleftharpoons P + E \tag{3.43}$$

involving an enzyme E as catalyst or an autocatalytic reaction of the type

$$A + X \rightleftharpoons 2X \quad . \tag{3.44}$$

Returning to the general form of a chemical reaction in (3.41), we write for the variation δN_i of the number of moles of X_i as caused by all reactions

$$\delta N_i = \sum_\rho \nu_{i\rho} \delta\xi_\rho \tag{3.45}$$

where

$$\nu_{i\rho} = - \nu_{i\rho}^f + \nu_{i\rho}^\rho \tag{3.46}$$

and ξ_ρ is the reaction coordinate which measures the molar displacement of the reaction. For a continuous reaction process, (3.45) takes the form

$$\dot{N}_i = \sum_\rho \nu_{i\rho} W_\rho \tag{3.47}$$

where $W_\rho = d\xi_\rho/dt$ is the molar rate of the reaction. The variation of the entropy δS associated with a variation $\delta\xi_\rho$ of the extents of the reactions is easily calculated as

$$\delta S = \sum_i \left(\frac{\partial S}{\partial N_i}\right)_{U,V,N_j} \delta N_i = - \frac{1}{T} \sum_i \mu_i \delta N_i = \frac{1}{T} \sum_\rho A_\rho \delta\xi_\rho \geq 0 \tag{3.48}$$

or

$$\dot{S} = \frac{1}{T} \sum_\rho A_\rho W_\rho \geq 0 \tag{3.49}$$

where

$$A_\rho = - \sum_i \nu_{i\rho} \mu_i \tag{3.50}$$

is the so-called affinity of reaction ρ.

From (3.48) or (3.49) we immediately conclude that the system is in equilibrium with respect to reaction ρ if the corresponding affinity A_ρ vanishes. For spontaneous evolutions towards the equilibrium we have

$$\dot{S} = \frac{1}{T} \sum_\rho A_\rho W_\rho > 0 \quad . \tag{3.51}$$

If only one single reaction occurs in the system, its rate W_ρ will have a direction from higher to lower values of its affinity whereas in the case of several simultaneous reactions it may happen that some of them are driven uphill by the others due to some unknown internal coupling as far as the inequality (3.51) remains satisfied.

Let us combine now the results of the present and the preceding section for a system which is in contact with a bath system with respect to an exchange of heat and matter and which simultaneously performs internal chemical reactions. Assuming that the temperature T of the system equals that of its bath system, which is a frequent situation in view of biological applications, the continuous entropy production $\dot{S}$ is then given as

$$\dot{S} = -\frac{1}{T}\sum_i \Delta\mu_i \cdot I_i + \frac{1}{T}\sum_\rho A_\rho W_\rho \geq 0 \tag{3.52}$$

where $\Delta\mu_i = \mu_i - \mu_{io}$ and μ_{io} is the chemical potential of component i in the bath system. Now it may happen that an external exchange drives a chemical reaction uphill or vice versa. The latter case is of particular interest in context with the phenomenon of the so-called active transport in biological membranes where ions are pumped against their concentration gradient by an appropriate chemical reaction. We should emphasize at this point that thermodynamics only gives the general framework for such possibilities. For a detailed description of active transport, one would have to construct a model which goes beyond the general theory. Thermodynamics, then, just tells us that such a model does not principally violate the basic laws. We shall discuss a simple model for active transport in Chapter 5 of this book.

3.7 Some Basic Relations of Equilibrium Thermodynamics

For later purposes we need a few further basic relations of equilibrium thermodynamics. Let us therefore restrict ourselves in this section to thermodynamic systems which are in external equilibrium with respect to a coupling to bath system as well as in internal equilibrium with respect to chemical reactions.

In (3.23) we have already introduced the concept of reversible changes of the entropy due to idealized reversible exchanges of heat as infinitely slow processes with negligible irreversible entropy production. This concept is immediately extended to include reversible changes of the entropy which may likewise be caused by infinitely slow exchanges of volume V and chemical components N_i. In the course of such processes, the system is not removed from equilibrium but is passing through a continuous series of equilibrium states. This means that in each state of such processes the system is characterized by its entropy S as a function of U, V and

N_1, N_2, ... Let dS be the reversible change of entropy as caused by such idealized reversible changes dU, dV, dN_i, i = 1, 2, ... of the macro-variables. dS may now be expressed as

$$dS = \left(\frac{\partial S}{\partial U}\right)_{V,N_i} dU + \left(\frac{\partial S}{\partial U}\right)_{U,N_i} dV + \sum_i \left(\frac{\partial S}{\partial N_i}\right)_{U,V,N_j} dN_j = \frac{1}{T}\, dU + \frac{p}{T}\, dV - \sum_i \frac{\mu_i}{T}\, dN_i \qquad (3.53)$$

where use has been made of the definitions of temperature (3.11), pressure (3.27), and chemical potentials (3.31). In writing (3.53), we have utilized the symbol d instead of δ in order to express that dS, dU, dV and dN_i are now total differentials of equilibrium state functions. From (3.53) we immediately recover the expressions (3.23) for the reversible change of entropy associated with a continuous reversible exchange of heat by setting $\dot{S}^{rev} = dS/dt$, $I_U^{rev} = dU/dt$ and $dV/dt = 0$, $dN_i/dt = 0$.

Equation (3.53) is known as Gibbs' fundamental relation. Transforming (3.53) into

$$dU = TdS - pdV + \sum_i \mu_i dN_i \qquad (3.54)$$

the internal energy U may now be interpreted as a function of S, V, N_i with a total differential expressed by (3.54). From (3.54) we immediately obtain equivalent definitions of the intensive variables T, p, μ_i, namely

$$T = \left(\frac{\partial U}{\partial S}\right)_{V,N_i}, \qquad p = -\left(\frac{\partial U}{\partial V}\right)_{S,N_i}, \qquad \mu_i = \left(\frac{\partial U}{\partial N_i}\right)_{S,V,N_j}. \qquad (3.55)$$

In (3.25) we have already introduced the free energy F of a thermodynamic system as F = U - TS (for equilibrium $T = T_0$). Taking the total differential of F we obtain by inserting dU from (3.54)

$$dF = dU - TdS - SdT = -SdT - pdV + \sum_i \mu_i dN_i \;. \qquad (3.56)$$

Eq. (3.56) expresses the change of the free energy F as caused by reversible changes of T, V and N_i. Eq. (3.56) leads us to interpret F as a function of T, V, N_i such that

$$S = -\left(\frac{\partial F}{\partial T}\right)_{V,N_i}, \qquad p = -\left(\frac{\partial F}{\partial V}\right)_{T,N_i}, \qquad \mu_i = \left(\frac{\partial F}{\partial N_i}\right)_{T,V,N_j}. \qquad (3.57)$$

The same mathematical transformation may be performed with the Gibbs' free energy G as defined in (3.30) to yield

$$dG = - SdT + Vdp + \sum_i \mu_i dN_i \tag{3.58}$$

and

$$S = -\left(\frac{\partial G}{\partial T}\right)_{p,N_i} , \qquad V = \left(\frac{\partial G}{\partial p}\right)_{T,N_i} , \qquad \mu_i = \left(\frac{\partial G}{\partial N_i}\right)_{T,p,N_j} . \tag{3.59}$$

Let us now return to the internal energy U as a function of S, V, N_1, N_2, ... : $U = U(S, V, N_1, N_2, \ldots)$. Recalling that both U and the independent variables S, V, N_1, N_2, ... are extensive macro-variables, we may write

$$\lambda U = U(\lambda S, \lambda V, \lambda N_1, \lambda N_2, \ldots) \tag{3.60}$$

which expresses the internal energy of a system with a λ-fold size compared to the original one. Taking the derivative with respect to λ of both sides of (3.60) and then setting $\lambda = 1$ we obtain

$$U = S\left(\frac{\partial U}{\partial S}\right)_{V,N_i} + V\left(\frac{\partial U}{\partial V}\right)_{S,N_i} + N_i\left(\frac{\partial U}{\partial N_i}\right)_{S,V,N_j} = TS - pV + \sum_i \mu_i N_i \tag{3.61}$$

where use has been made of (3.55). Eq. (3.61) has some important implications. First we observe that due to (3.61) the Gibbs' free energy G may be expressed as

$$G = U - TS + pV = \sum_i \mu_i N_i \tag{3.62}$$

and the function ψ introduced in (3.40) is obtained as

$$\psi = U - TS - \sum_i \mu_i N_i = - pV . \tag{3.63}$$

A second consequence of (3.61) is derived by taking the total differential of both sides of (3.61) and inserting dU from (3.54):

$$SdT - Vdp + \sum_i N_i d\mu_i = 0 . \tag{3.64}$$

Eq. (3.64) is the so-called Gibbs-Duhem relation. It expresses the fact that the intensive variables T, p, μ_1, μ_2, ... are not independent of each other but bound by exactly one differential relation.

Finally, (3.61) allows us to rewrite Gibbs' relation (3.54) in terms of volume densities $u = U/V$, $s = S/V$, $c_i = N_i/V$ which upon insertion into (3.54) together with (3.61) immediately yields

$$du = Tds + \sum_i \mu_i dc_i \tag{3.65}$$

and similar expressions for $f = F/V$ and $g = G/V$.

3.8 Perfect Gases and Ideal Mixtures

The relations which we have derived in the preceding section give a general framework for the equilibrium properties of thermodynamic system. The individual properties of particular systems have to be defined by so-called equations of state. An important class of systems is that of perfect gases with an equation of state given as

$$pV = NRT \tag{3.66}$$

where N is the number of moles for the case that only one kind of molecule is present in the system and R is the so-called gas constant with a value $R \approx 2$ cal/K mole independent of the kind of molecules. An important consequence from (3.66) is obtained by making use of the relation

$$\left(\frac{\partial U}{\partial V}\right)_{T,N_i} = T\left(\frac{\partial p}{\partial T}\right)_{V,N_i} - p \tag{3.67}$$

cf. problem 3.

Inserting (3.66) into (3.67) we directly obtain $(\partial U/\partial V)_{T,N} = 0$ such that the internal energy U for given T and N is independent of V: $U = U(T,N)$. Since both U and N are extensive whereas T is intensive, $U(T,N)$ must be of the form

$$U = N\tilde{u}(T) \tag{3.68}$$

where $\tilde{u}(T)$ is the molar internal energy. Introducing now also the molar entropy $\tilde{s}$ and the molar volume $\tilde{v}$ by $\tilde{s} = S/N$ and $\tilde{v} = V/N$ we may write the Gibbs' relation (3.53) for one mole of the perfect gas ($N = 1$ fixed) as

$$d\tilde{s} = \frac{d\tilde{u}}{T} + \frac{p}{T}\,d\tilde{v} = \frac{c(T)}{T}\,dT + \frac{R}{\tilde{v}}\,d\tilde{v} \tag{3.69}$$

where $c(T) = d\tilde{u}(T)/dT$ is the molar heat capacity (at constant volume). Since (3.69) expresses $d\tilde{s}$ as a sum of two terms each of which depends only on one of the variables T and $\tilde{v}$ we can directly integrate (3.69) to obtain

$$\tilde{s}(T,\tilde{v}) = \tilde{s}(T,\tilde{v} = 1) + R \ln \tilde{v} \quad . \tag{3.70}$$

$\tilde{s}(T,\tilde{v} = 1)$ is a function of T only and expresses the value of the molar entropy at unit molar volume $\tilde{v} = 1$. Clearly, it depends on the choice of units for $\tilde{v}$. From (3.70) we can easily derive an expression for the molar Gibbs' free energy $\tilde{g} = G/N = \tilde{u} - T\tilde{s} + p\tilde{v}$ which due to (3.62) coincides with the chemical potential for a one-component gas such that eventually

$$\mu(T,\tilde{v}) = \mu(T,\tilde{v} = 1) - RT \ln \tilde{v} \tag{3.71}$$

or

$$\mu(T,p) = \mu(T,p = 1) + RT \ln p \tag{3.72}$$

where again use has been made of the equation of state, (3.66).

If in a mixture of perfect gases with mole numbers N_1, N_2, ... each component behaves as if it were independent of the others, the equation of state of the component i is given as

$$p_i V = N_i RT \qquad i = 1, 2, \ldots \tag{3.73}$$

where p_i is the so-called partial pressure of component i. The total pressure p of the mixture is the sum of all p_i. In analogy to (3.72) the chemical potential μ_i of component i is now given as

$$\mu_i(T,p_i) = \mu_i(T,p_i = 1) + RT \ln p_i \tag{3.74}$$

or, upon inserting (3.73), as

$$\mu_i(T,c_i) = \mu_i(T,c_i = 1) + RT \ln c_i \tag{3.75}$$

where $c_i = N_i/V$ is the molar concentration of components i. Another way of representing μ_i is obtained by introducing the so-called mole fractions $x_i = N_i/N$ where $N = N_1 + N_2 + \ldots$ is the total mole number. Expressing c_i in (3.75) by x_i and making use of (3.73) one obtains

$$\mu_i(T,p,x_i) = \tilde{\mu}_i^0(T,p) + RT \ln x_i \tag{3.76}$$

where $\tilde{\mu}_i^0(T,p)$ is the so-called reference potential.

The validity of (3.76) actually goes far beyond mixtures of perfect gases. Systems in which the chemical potentials of the components can be expressed by (3.76) are called ideal systems. A special case of ideal systems are dilute solutions. A statistical derivation of (3.76) for dilute solutions may be found in LANDAU-LIFSHITZ, Theoretical Physics, Vol. V (1968). In dilute solutions, the mole fraction of a solute is approximately given as $x_i = N_i/N_0$ where N_0 is the number of moles of the solvent. This enables us to rewrite the chemical potential of a solute approximately in terms of its concentration c_i as

$$\mu_i \approx \mu_i^0(T,p) + RT \ln c_i \tag{3.77}$$

where the term $RT \ln(V/N_0)$ has been expressed by T and p and incorporated in $\mu_i^0(T,p)$. We have made use of (3.77) already in earlier sections, cf. (2.18) or (2.44).

3.9 Linear Irreversible Thermodynamics

In Sections 3.4, 3.5 and 3.6 we have derived expressions for the entropy production associated with spontaneous evolutions in systems which are not in equilibrium states. Inspection of these expressions as given in (3.19), (3.29), (3.33) and (3.49) shows us that the most general case of entropy production may be written in the form

$$\dot{S}^{(i)} = \sum_\alpha F_\alpha \cdot I_\alpha \geq 0 \tag{3.78}$$

where F_α and I_α are conjugated pairs of so-called generalized thermodynamic forces and fluxes, respectively, given by

$$\left.\begin{array}{lllll} F_\alpha : & \Delta \frac{1}{T} , & \Delta \frac{p}{T} , & - \Delta \frac{\mu_i}{T} , & A_\rho/T \\ I_\alpha : & I_U , & I_V , & I_i , & W_\rho \end{array}\right\} . \tag{3.79}$$

The summation index α in (3.78) runs through all irreversible processes in the system, and the superscript (i) in $\dot{S}^{(i)}$ indicates that only the irreversible part of the change of entropy is included.

From (3.79) we see that the forces F_α are differences of intensive variables. This also applies to chemical reactions since according to (3.50) the affinity A_ρ is defined as a balance of chemical potentials weighted by the stoichiometric coefficients. On the other hand, the fluxes in (3.79) are time derivatives of extensive variables.

Clearly, the system is in complete equilibrium only if all of the forces vanish, $F_\alpha = 0$, $\alpha = 1,2, \ldots$ which implies that also all of the fluxes vanish, $I_\alpha = 0$, $\alpha = 1,2, \ldots$ and vice versa. This conclusion suggests considering the forces F_α as causing the fluxes I_α as their effects or vice versa, that is, to write

$$I_\alpha = I_\alpha(F_1, F_2, F_3, \ldots) \quad . \tag{3.80}$$

Two important points should be emphasized in context with (3.80). First, it is usually observed in nonequilibrium systems that a force F_α not only gives rise to its conjugated flux I_α, but also to other fluxes I_β with $\beta \neq \alpha$. This is expressed in (3.80) by writing the flux I_α as a function of all forces F_1, F_2, ... The second point is that (3.80) is an incomplete way of writing the mathematical relation between forces and fluxes: a flux I_α will quite generally depend not only on all forces, but also on the parameters of the equilibrium state, the so-called reference state, relative to which the differences in the F_α are defined. For a detailed discussion of this latter point, the reader is referred to the review article by SAUER (1973).

The right-hand side of (3.80) can be expanded in a Taylor series in powers of the forces F_1, F_2 .... The coefficients of this expansion will depend on the parameters of the reference state. Since all $I_\alpha = 0$ if all $F_\alpha = 0$, there will be no zero-order terms. Truncating the Taylor series after the first-order term, we get a linear relation of the type

$$I_\alpha = \sum_\beta L_{\alpha\beta} F_\beta \tag{3.81}$$

where the $L_{\alpha\beta}$ are the so-called phenomenological coefficients. Clearly (3.81) will be a sufficient approximation only for small forces F_α and fluxes I_α. The range of validity of (3.81) defines the so-called linear irreversible thermodynamics (LIT). For quite a lot of irreversible phenomena LIT is a really good approximation. As an example for a linear relation of the LIT type in (3.81) cf. Fick's law (2.16) or Ohm's law (2.17) for diffusion where $L = D\bar{c}/\mathcal{R}a = T/R$. For chemical reactions and in particular for biological processes, however, LIT is quite insufficient.

Inserting (3.81) into (3.78) we obtain

$$\dot{S}^{(i)} = \sum_{\alpha,\beta} L_{\alpha\beta} F_\alpha F_\beta \geq 0 \tag{3.82}$$

which expresses the entropy production as a nonnegative quadratic form in terms of the forces F_α.

A very important property of the linear phenomenological coefficients is Onsager's reciprocity relation

$$L_{\alpha\beta} = L_{\beta\alpha} \tag{3.83}$$

which can be proved under rather general statistical conditions for the system. For a detailed proof, the reader is referred to the monograph by de GROOT and MAZUR (1962).

Let us conclude our brief outline of LIT by asking what happens to a system if it is prevented from attaining an equilibrium state, for example by fixing the values of some of the forces, say $F_1, F_2, \ldots F_\nu$. Since now the system cannot decrease its entropy production completely down to $\dot{S}^{(i)} = 0$, we would intuitively guess that it will try to diminish the value of $\dot{S}^{(i)}$ at least as far as possible. This is exactly what happens within LIT. The condition of minimal $\dot{S}^{(i)}$ is obtained by setting the partial derivatives of $\dot{S}^{(i)}$ with respect to the unrestricted forces $F_{\nu+1}, F_{\nu+2}, \ldots$ equal to zero:

$$\begin{aligned}\frac{\partial \dot{S}^{(i)}}{\partial F_{\nu+\alpha}} &= \sum_\beta \left(L_{\nu+\alpha,\beta} + L_{\beta,\nu+\alpha}\right) F_\beta = \\ &= 2 \sum_\beta L_{\nu+\alpha,\beta} F_\beta = 2\, I_{\nu+\alpha} = 0\end{aligned} \tag{3.84}$$

where use has been made of Onsager's reciprocity relations (3.83). The condition $I_{\nu+\alpha} = 0$ means that the system is in a stationary state with respect to the unrestricted extensive variables incorporated in $I_{\nu+\alpha}$. Hence, the stationary state at which the system will eventually arrive, is a state of minimal entropy production. The fact that $\dot{S}^{(i)}$ is really minimal and not maximal is a consequence of the property that $\dot{S}^{(i)}$ is a nonnegative quadratic form.

The intrinsic tendency of the system to decrease its entropy production as far as possible is known as the principle of minimum entropy production. It defines a general direction of spontaneous evolutions in thermodynamic systems which are not in equilibrium states. The proof of this principle depends on the validity of Onsager's symmetry relation. It is thus restricted to the range of validity of LIT. Outside of this range, the situation may change drastically. In fact, we have been concerned with such a case already in Section 2.7 where in the model for the control of metabolic reaction chains, a situation far from the equilibrium and hence far from LIT does not lead to a steady state of minimal entropy production but to a limit cycle. We shall study similar effects particularly in view of biological applications in Chapter 6.

Problems

1) Derive (3.28) and prove $\delta G = - T_o \delta S_{tot}$ for a system in heat and mechanical contact with a bath system where G is defined by (3.30) and δS_{tot} is the change of the total entropy of the system and its bath. This then also proves that G becomes minimal in the equilibrium.
2) Derive Gibbs' relation for the volume densities $f = F/V$ and $g = G/V$ of the free energy F and Gibbs' free energy G.
3) Prove (3.67) by deriving

$$\left(\frac{\partial U}{\partial V}\right)_{T,N_i} = T\left(\frac{\partial S}{\partial V}\right)_{T,N_i} - p$$

from (3.54) and

$$\left(\frac{\partial S}{\partial V}\right)_{T,N_i} = \left(\frac{\partial p}{\partial T}\right)_{V,N_i}$$

from (3.56).
4) Derive the mass action law for ideal systems,

$$\prod_i (x_i)^{\nu_{i\rho}} = K(T,p) = \exp\left\{-\frac{1}{RT}\sum_i \nu_{i\rho}\mu_i^o(T,p)\right\}$$

for $\rho = 1,2, \ldots$ from the equilibrium condition $A_\rho = 0$. Convince yourself that the enzyme E which acts as a catalyst in the Michaelis-Menten model of (2.1) does not influence the equilibrium balance. Show that the steady state condition (2.6) reduces to the mass action law if $\bar{J} = 0$.
5) Rewrite (3.81) for the diffusion of an ion under the simultaneous influence of differences of a chemical and electrical potential into the original form of Ohm's law for the electric flux I_e and show $I_e = V/R_e$ where V is an effective voltage difference and $R_e = T/(z^2e^2F^2L)$.

6) Calculate the phenomenological coefficient L for a chemical reaction as given in (3.41) by making a product ansatz for the reaction flux like that in (2.4) and assuming that the reaction takes place in a dilute solution.

4. Networks

4.1 Network Language and Its Processes

In this chapter, we shall introduce a network language in order to represent the thermodynamic processes and their interactions which typically occur in biological systems. The question whether there is a real need for such a new language is primarily a question of convenience. After having introduced the physical basis of thermodynamic processes like transfer of heat and matter or chemical reactions in the previous chapter, the introduction of a new language does not bring into play any substantially new physical aspect. On the other hand, the utilization of an appropriate language is very often a decisive step when formulating and solving a problem.

The application of a network language to the dynamic modelling of biophysical systems was suggested first by OSTER, PERELSON and A. KATCHALSKY (1973). Previously, the relation between network theory and irreversible thermodynamics had been already established by MEIXNER (1966). The rules of the network language are such that the thermodynamic condition for the processes formulated in this language are automatically satisfied. Moreover, the network method is a typical system-analytic method in that the development of a model for a certain biological phenomenon starts from a purely phenomenological point of view. Only those processes and interactions will be displayed by the network model which are believed to be essential for the phenomenon in view whereas all structural and functional details which do not directly interfere with the phenomenon are suppressed. On the other hand, the elements of the network language are such that at each state during the course of a proceding network analysis a direct molecular interpretation of the model and the processes involved in it can be given. A more detailed introduction into the technique of network representation with particular regard to applications in engineering sciences is found in the book by THOMA (1975).

In Sections 2.3 and 2.4 we have already utilized network representations in order to express diffusion of molecules and ions across membranes and their storage in the interior of the membrane, in the case of relaxation phenomena. The network language to be developed in this chapter will also be appropriate to represent all other models discussed in Chapter 2 and even further models which will be developed in Chapters 5 and 6. The basic processes involved in those models are diffusion, reaction and storage. In the thermodynamic considerations of Chapter 3 we have learned that diffusion, i.e., exchange of matter, and reaction are dissipative processes characterized by a nonvanishing irreversible entropy production. In contrast, storage of matter or charge will turn out to be a reversible or equilibrium process which is expressed by a generalized capacitance to be described in the following section. The elements for the dissipative processes of diffusion and reaction will be introduced in Sections 4.4, 4.5 and 4.6.

Throughout the investigations in this book the phenomenon of volume flow or convection will be excluded by assuming vanishing differences of hydrodynamic pressure in the systems. This is an incisive restriction, and a prominent biological counterexample is the flow of blood through the arteries and veins. Although the network language is capable of including such convection processes without any principal difficulty, we shall restrict ourselves to diffusion as far as transport of matter in concerned since inclusion of convection would require a discussion of the basic hydrodynamic background and thus go beyond the framework of an introductory text.

4.2 Storage Elements: Generalized Capacitances

As a generalized capacitance we simply define any extensive quantity E_i of the system which may change its value with time t: $E_i = E_i(t)$. In most cases, the E_i will be the mole numbers of chemical species, $E_i \triangleq N_i$, or the electric charge $E_i \triangleq Q$. The time variation of E_i is due to fluxes I_i in the system: $I_i = dE_i/dt$ which defines a capacitive flux I_i. Frequently, it is more convenient to define a flux not by the time change of E_i but by that of its volume density or concentration $c_i =$ $= E_i/V$, that is to write

$$\frac{dc_i}{dt} = J_i \quad , \qquad X_i \xleftarrow[J_i]{c_i} \tag{4.1}$$

where X_i is the symbol for the variable with concentration c_i. The left-hand side of (4.1) simply defines the flux J_i, and the right-hand side is the network symbol for this definition. The direction of the arrow represents a reference direction for the flux, i.e., the direction in which the flux is counted positive. The reference direction of (4.1) is the so-called associated reference direction. We also could have defined

$$\frac{dc_i}{dt} = - J_i' \qquad X_i \xrightarrow[J_i']{c_i} \tag{4.2}$$

which is the so-called nonassociated reference direction. It is clear that the fluxes J_i need not be constant but may themselves depend on time.

The reason for calling a time-dependent concentration a capacitance has already been explained in Section 2.3 in context with the relaxation behaviour of the black-box membrane model. Let us assume that the concentration c_i of a molecule of kind i is given as a unique function of its chemical potential μ_i : $c_i = c_i(\mu_i)$. Taking the time derivative and making use of the associated reference direction we then obtain

$$J_i = \frac{dc_i}{dt} = \frac{dc_i}{d\mu_i}\frac{d\mu_i}{dt} = C_i(\mu_i)\,\frac{d\mu_i}{dt} \tag{4.3}$$

where

$$C_i(\mu_i) = \frac{dc_i}{d\mu_i} \quad . \tag{4.4}$$

The right-hand side of (4.3) precisely has the form of the constitutive relation for an electrical capacitance. We thus call $C_i(\mu_i)$ as defined in (4.4) a material capacity. For ideal systems like dilute solutions we obtain by inserting (3.77) into (4.4)

$$C_i(\mu_i) = \frac{1}{RT}\exp\frac{\mu_i - \mu_i^o}{RT} = \frac{c_i}{RT} \tag{4.5}$$

in agreement with (2.25). (The extra factor a in (2.25) stems from the convention of counting the material capacity of a membrane per area instead of per volume as in (4.4).

Not only (4.5) but even the assumption that c_i is a unique function of its chemical potential μ_i are approximations which are valid only for special systems like ideal systems. In the general case, the concentration c_i of molecules of kind i will depend on the chemical potentials μ_j of all other molecules j including i. In this situation we have

$$\frac{dc_i}{dt} = \sum_j C_{ij}\,\frac{d\mu_j}{dt} = J_i \tag{4.6}$$

where C_{ij} is the capacitance matrix defined by

$$C_{ij} = \left(\frac{\partial c_i}{\partial \mu_j}\right)_{T,\mu_k} \quad . \tag{4.7}$$

We now have two possibilities to confine the meaning of generalized capacitances depending on the choice of variables:

a) a capacitance relates fluxes and potentials to each other such that (4.6) becomes its constitutive relation and the network symbolization has to take into account the fact that now all species of molecules are coupled by the matrix C_{ij}:

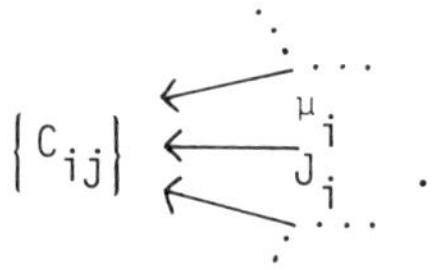

(4.8)

In this version, the capacitance becomes a so-called n-port element if there are n species of molecules present in the system, each port being characterized by a pair of potential and flux variables (μ_i, J_i).

b) As in (4.1), a capacitance relates fluxes and concentrations to each other such that there is a separate one-port capacitance for each of the n species of molecules characterized by the corresponding pair (c_i, J_i) of concentration and flux variables.

There are numerous points of view which are in favour of one of the above possibilities. The choice of (μ_i, J_i) as so-called conjugated variables is supported by the fact that the product μ_i, J_i has the dimension of a power rate and is closely related to the time change of the energy of the capacitance as we shall see in Chapter 7. Generally speaking, the potential-flux representation (μ_i, J_i) of the capacitances is the appropriate language for a thermodynamic analysis of the system. Nevertheless, it is often more convenient to make use of the concentration-flux representation (c_i, J_i) particularly in cases where the concentration-potential relation is not known and one wants to represent the system by a network which is topologically as simple as possible and to avoid multi-port elements of the type of (4.8). For the following considerations in this book, we shall make use of both representations depending on the particular problem we have in view.

Let us conclude the consideration of generalized capacitances with the important remark that a capacitance represents a thermodynamically reversible element. Quite formally, this follows from the fact that the time changes of the concentrations $\dot{c}_i = dc_i/dt$ or the fluxes J_i in (4.1) and (4.2) may have any sign as to be distinguished from irreversible processes like transfers of energy, volume and matter or chemical reactions as discussed in Sections 3.4 to 3.6. The latter processes involve the production of entropy and thus all time changes associated with them are such that this production is nonnegative.

4.3 Kirchhoff's Current Law and the 0-Junction

In many respects, the leading ideas for the conventions of the generalized thermodynamic network language are the rules of the so-called linear networks of electrical engineering. Before introducing a new element in this section, let us therefore establish the relationship between a generalized capacitance as introduced in the so-called bond-graph representation in (4.1) and a material capacitance as considered in Section 2.3, Fig. 3, for the linear network of the black-box model for transport across membranes. The translation rule we are looking for roughly says that all ground connections of elements in linear networks are to be cancelled in order to construct the bond-graph representation. Clearly, this is a unique and consistent simplification procedure if for all elements which have ground connections a common reference potential is defined. We shall come back to this translation rule in more detail when transforming the network of Fig. 3 stepwise into the bond-graph language. In the present context, we just keep in mind the capacitance equivalence

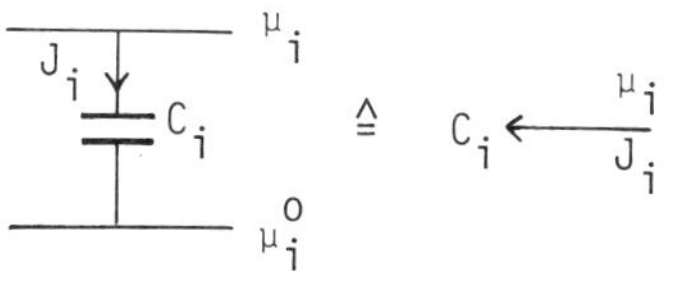

(4.9)

where μ_i^0 has been chosen as the common ground potential.

As in the example of Fig. 3, capacitances are very often charged or discharged not only by just one single flux but by a net flux resulting from different connections between the capacitance and other elements in the network. For this purpose, we need a connection element in the bond-graph network language which is equivalent to the soldering point of linear networks and characterized by the property to sum fluxes at given and constant values of the concentration. This element is called a 0-junction and defined by

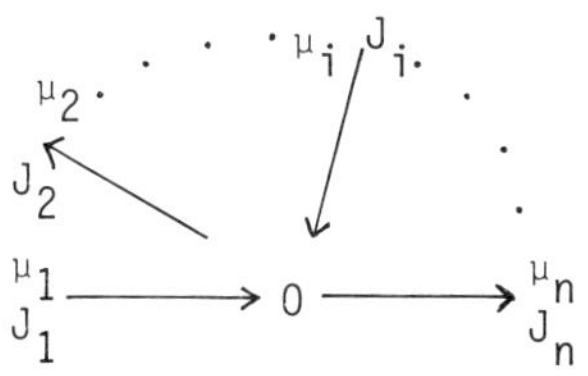

(4.10)

where

$$\sum_i \sigma_i J_i = 0 \tag{4.11}$$

and

$$\mu_1 = \mu_2 = \ldots = \mu_n \quad \text{or} \quad c_1 = c_2 = \ldots = c_n \; . \tag{4.12}$$

In (4.11) $\sigma_i = +1$ or $\sigma_i = -1$ depending on whether the reference orientation of J_i is directed out of or into the 0-junction.

It is evident that (4.10) is the bond-graph version of the soldering point of linear networks and thus (4.11) is nothing else than Kirchhoff's current law, in the following abbreviated as KCL.

4.4 Unimolecular Reactions

The bond-graph element for unimolecular reactions which we shall introduce in this section turns out to be the essential dissipative element of all types of networks to be discussed in the following chapters. It not only represents the basic description for unimolecular reactions but also for reactions of arbitrary order and for diffusion as well. Our starting point is an ansatz for the reaction flux J of an unimolecular reaction $X_1 \rightleftharpoons X_2$.

$$J = kc_1 - k'c_2 \tag{4.13}$$

where c_1 and c_2 are the concentrations of the reactants X_1 and X_2 respectively, and J counts the net rate of moles per second converted from X_1 to X_2 or vice versa with reference orientation from X_1 to X_2. The k and k' are the forward and reverse reaction rate constants, respectively.

The ansatz (4.13) corresponds to that of (2.4) for the enzyme reaction system (2.1) in Section 2.2. There we have argued already that the derivation of an ansatz like (4.13) or (2.4) is far from being trivial although its range of validity is surprisingly wide including even such complex processes as in the population dynamic model of Section 2.6.

Returning to (4.13) we now represent this process by a 2-port bond-graph element which reads

$$\xrightarrow[J]{c_1} R \xrightarrow[J]{c_2} . \tag{4.14}$$

The arrows of the port bonds of the element R in (4.14) denote the reference orientation for the flux J, and (4.13) is the constitutive relation of the element R. This relation includes the condition that the fluxes along the two port bonds are equal.

Combining now the port bonds of the reaction element R with that of material capacitances for the molecules X_1 and X_2 we obtain our first simple complete network:

$$X_1 \xrightarrow[J]{c_1} R \xrightarrow[J]{c_2} X_2 . \tag{4.15}$$

According to the constitutive relations (4.1) and (4.2) for the capacitances the time changes of c_1 and c_2 are given as

$$\frac{dc_1}{dt} = - J \quad , \quad \frac{dc_2}{dt} = + J \quad . \tag{4.16}$$

From (4.16) we see that the sum of c_1 and c_2 is conserved: $c_1 + c_2 = c = \text{const}$. Inserting (4.13) into (4.16) and replacing c_2 by $c - c_1$ gives

$$\frac{dc_1}{dt} = - (k + k')c_1 + k'c \tag{4.17}$$

which upon integrating yields

$$c_1(t) = \frac{k'}{k + k'} c + \left(c_1(0) - \frac{k'}{k + k'} c\right) e^{-(k + k')t} \quad . \tag{4.18}$$

Since the system (4.15) is materially closed, the asymptotic values of c_1 and c_2 obtained from (4.18) for $t \longrightarrow \infty$ are the equilibrium concentrations

$$\bar{c}_1 = k'c/(k + k') \quad , \quad \bar{c}_2 = kc/(k + k') \quad , \quad \bar{c}_1/\bar{c}_2 = k'/k \quad . \tag{4.19}$$

In (4.14) we have represented the R element in the concentration-flux language. In the potential-flux language we would have to replace the concentrations c_1, c_2 by the corresponding potentials μ_1, μ_2, respectively. For ideal systems, cf. (3.77), the reaction flux J would then be given by

$$J = k \exp \frac{\mu_1 - \mu_1^o}{RT} - k' \exp \frac{\mu_2 - \mu_2^o}{RT} \tag{4.20}$$

where μ_1^o, μ_2^o are the reference potentials of X_1, X_2 extrapolated at $c_1 = 1$, $c_2 = 1$. In (3.50) of Section 3.6 we have introduced the affinity of a reaction which in our simple example reads

$$A = \mu_1 - \mu_2 \quad . \tag{4.21}$$

For vanishing affinity, A = 0, the system must come to equilibrium, i.e., J = 0. Inserting this condition into (4.20) yields

$$k \exp\left(-\frac{\mu_1^o}{RT}\right) = k' \exp\left(-\frac{\mu_2^o}{RT}\right) \equiv \kappa \quad . \tag{4.22}$$

With this definition of κ we can rewrite (4.20) as

$$J = \kappa \left\{ \exp \frac{A^f}{RT} - \exp \frac{A^r}{RT} \right\} \tag{4.23}$$

where the so-called forward and reverse infinities A^f and A^r have been introduced as

$$A^f = \mu_1 \quad , \qquad A^r = \mu_2 \quad , \qquad A = A^f - A^r \quad . \tag{4.24}$$

We see that the reaction flux J cannot be represented as a unique function of its conjugated A but depends separately on A^f and A^r. Of course, one could have written J as a function of A, for example

$$J = \kappa \exp \frac{\mu_1}{RT} \cdot \left(\exp \frac{A}{RT} - 1 \right) \tag{4.25}$$

but then J also depends on μ_1 which plays the role of a reference state for the definition of A as mentioned already in context with (3.80) in Section 3.9.

4.5 Higher-Order Reactions, Kirchhoff's Voltage Law and the 1-Junction

In this section, we shall extend the use of the reaction element R introduced in Section 4.4 to include also higher-order chemical reactions. Let us consider a reaction of the type $X_1 + X_2 \rightleftharpoons X_3$ with the usual product ansatz for its flux given by

$$J = kc_1c_2 - k'c_3 \quad . \tag{4.26}$$

A bond-graph network representation of such a reaction evidently has to involve an element for the multiplication of the concentrations c_1 and c_2 disposed by their corresponding capacitances. On the other hand, the reaction (4.26) consumes or produces equal rates of X_1 and X_2 such that the required element performs the multiplication at equal fluxes. We call this element a 1-junction. Its representation in the reaction of the above example is given by

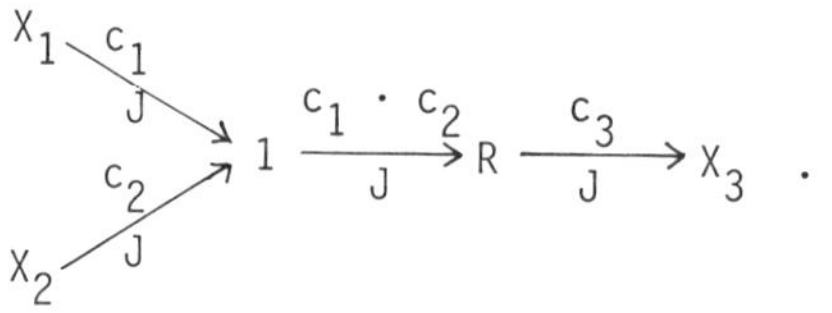

(4.27)

In the potential-flux language of ideal systems, multiplication of concentrations is equivalent to summation of the corresponding potentials since the potentials depend logarithmically on the concentrations. We therefore generalize the definition of the 1-junction in the potential-flux language as

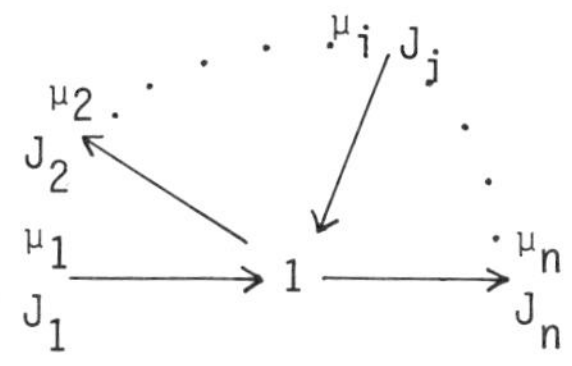

(4.28)

where

$$\sum_i \sigma_i \mu_i = 0 \tag{4.29}$$

$$J_1 = J_2 = \dots = J_n \tag{4.30}$$

and $\sigma_i = +1$ or $\sigma_i = -1$ depending on whether the reference orientation of J_i is directed out of or into the 1-junction. This definition is very similar to that of the 0-junction in Section 4.3, the only difference lying in the interchanged roles of fluxes and potentials. From this analogy we also conclude that (4.29) now expresses Kirchhoff's voltage law, abbreviated as KVL, as (4.11) expressed KCL. In linear networks, KCL is a condition for junctions whereas KVL is a condition for meshes. In bond-graph networks, KCL and KVL are conditions for two different types of junctions.

For ideal systems, we can immediately retranslate the above definition of 1-junction into the concentration-flux language to obtain from (4.29)

$$\prod_i (c_i)^{\sigma_i} = 1 \tag{4.31}$$

whereas (4.30) remains unchanged.

We are now also in a position to construct the network for the reaction $2X_1 \rightleftharpoons X_2$. According to our above rules this network is given in the potential-flux language as

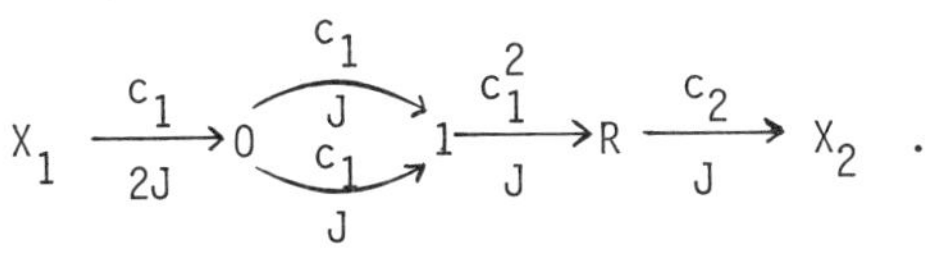

(4.32)

The effect of the combination of 0- and 1-junctions in (4.32) is a transformation from $(c_1, 2J)$ to (c_1^2, J). For such a transformation, we now introduce a short-writing element, the so-called transducer TD defined as

$$\xrightarrow[J]{c} \underset{(r)}{TD} \xrightarrow{c'}_{J'} \qquad (4.33)$$

$$c' = (c)^r \qquad J' = \frac{1}{r} J \ . \qquad (4.34)$$

The positive number r which in general need not be an integer is called the modulus of the transducer. For the example in (4.32) we have r = 2. Making use of the transducer, (4.32) can now be rewritten as

$$X_1 \xrightarrow[2J]{c_1} \underset{(2)}{TD} \xrightarrow[J]{c_1^2} R \xrightarrow[J]{c_2} X_2 \ . \qquad (4.35)$$

We see from (4.35) that the transducer serves to represent stoichiometric coefficients unequal to one in chemical reactions. Beyond this role, it will also turn out to be a coupling element for different fluxes as we shall see in the following chapter. Let us also notice that in the potential-flux language the transducer is definded by the relations

$$\xrightarrow[J]{\mu} \underset{(r)}{TD} \xrightarrow{\mu'}_{J'} \qquad (4.36)$$

$$\mu' = r\mu \qquad J' = \frac{1}{r} J \qquad (4.37)$$

which replace (4.33) and (4.34). From (4.37) we see that the transducer is a power conserving element since $\mu J = \mu' J'$. From (4.34) or (4.47) we also derive that the inversion of the reference orientations for the transducer is equivalent to inverting its modulus: $r \rightarrow 1/r$.

Let us conclude this section by translating the model for the enzyme-catalyzed reaction of Section 2.2 into the bond-graph language. According to the rules we have developed so far the corresponding network reads in the concentration-flux language

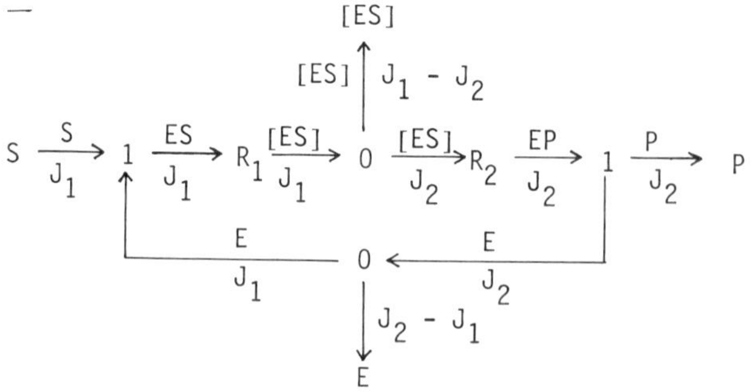

$$
\begin{array}{c}
\text{[ES]} \\
\text{[ES]} \Big\uparrow J_1 - J_2 \\
S \xrightarrow[J_1]{S} 1 \xrightarrow[J_1]{ES} R_1 \xrightarrow[J_1]{\text{[ES]}} 0 \xrightarrow[J_2]{\text{[ES]}} R_2 \xrightarrow[J_2]{EP} 1 \xrightarrow[J_2]{P} P \\
\underset{J_1}{E} \qquad \quad 0 \xleftarrow{} \underset{J_2}{E} \\
\Big\downarrow J_2 - J_1 \\
E
\end{array}
\qquad (4.38)
$$

where we have chosen the values of concentrations and fluxes such that the relations implied by the 0- and 1-junctions are automatically satisfied. Eqs. (2.2) and (2.4) of Section 2.2 are now the constitutive relations for the capacitances of the enzyme E and the enzyme-substrate complex [ES] and for the reactions R_1 and R_2. Regarding the constitutive relations for S and P, we have assumed in Section 2.2 that the concentrations of S and P are kept constant. This assumption is evidently equivalent to saying that the capacitances of S and P are infinitely large such that J_1 and J_2, although nonvanishing, will not influence the concentrations of S and P.

4.6 Diffusion

As in the case of chemical reactions in Sections 4.4 and 4.5 the starting point for the bond-graph representation of diffusion processes is an ansatz for the flux of this process. Let us consider the diffusion flux J of a neutral molecule under the influence of a space-dependent concentration c of this molecule and let us simplify this situation by assuming that the concentration only depends on one single spatial direction, say on the coordinate x: $c = c(x)$. Then Fick's first law gives the diffusional flux parallel to the x-direction as

$$J = -D \frac{dc(x)}{dx} \tag{4.39}$$

where D is a constant, the so-called diffusion coefficient, and J is measured as the number of moles transported per units of area and time.

The formulation (4.39) is suitable for systems where the concentration c is a continuous function of the coordinate x. Now, the most frequent cases, particularly in the biological field, are discrete systems consisting of nearly homogeneous subsystems with different concentrations of all kinds of molecules. For such systems (4.39) is replaced by the corresponding discrete formulation

$$J = \alpha(c - c') \tag{4.40}$$

where now J is the diffusional flux from subsystem with concentration c to subsystem with concentration c'. The formulation (4.40) applies to the special case that thermodynamic equilibrium between the subsystems with respect to diffusion is present if the concentrations are equal: $c = c'$. Since this need not be the case if the physical and chemical structures of the subsystems are different, the most general ansatz for the diffusional flux between the subsystems is

$$J = \alpha c - \alpha' c' \quad . \tag{4.41}$$

Now the equilibrium partition of the diffusing molecule between the subsystems is given as $c/c' = \alpha'/\alpha$. Eq. (4.41) expresses the diffusion flux in discrete systems in a form which is identical to that of the reaction flux (4.13) of a unimolecular reaction. Consequently, we chose the same bond-graph element (4.14) for representing diffusion flux as for representing reaction flux.

With this convention, we should now be ready to construct the bond-graph network of the black-box model for transport across membranes in Section 2.3. Since we have formulated this model and its representative linear network in Fig. 2 in terms of chemical potentials, we rewrite (4.41) in terms of the corresponding potentials μ, μ' quite similarly as (4.23)

$$J = \kappa \left(\exp \frac{\mu}{RT} - \exp \frac{\mu'}{RT} \right) \quad . \tag{4.42}$$

For small values of the difference $\Delta\mu = \mu - \mu'$ we expand J up to the first-order term in $\Delta\mu$. Referring to the explicit calculation of Section 2.3, (2.19), we obtain

$$J = \frac{1}{R} \Delta\mu \ , \qquad R = \frac{RT}{\kappa \cdot \bar{c}} \tag{4.43}$$

(In (2.19) the constant κ has been expressed as D/a). In the linear networks of Figs. 2 and 3, the material resistance R has been represented by a usual 2-pole Ohmic resistance. According to the translation rules between linear and bond-graph networks which we have established already, this means that the bond-graph version of the material resistance is a one-port element. This is easily seen when interpreting the relation $\Delta\mu - \mu + \mu' = 0$ as a KVL-relation (4.29) for potentials such that the bond-graph representation for the material resistor can be written as

$$\begin{array}{ccccc} & & R & & \\ & & \Delta\mu \uparrow J & & \\ \xrightarrow[J]{\mu} & & 1 & & \xrightarrow[J]{\mu'} \end{array} \quad . \tag{4.44}$$

Now, $J = \Delta\mu/R$ becomes the constitutive relation for the one-port element R in (4.44).

In combination with the conventions from Section 4.2 concerning the bond-graph representation of material capacitances, we can now translate the black-box model Fig. 3 in Section 2.3 for transport across membranes including relaxation processes into the bond-graph language as

$$\begin{array}{ccccccc} & R/2 & & C_m & & R/2 & \\ & \Delta\mu \uparrow J & & \mu_m \uparrow J_m & & \Delta\mu' \uparrow J' & \\ \xrightarrow[J]{\mu} & & \xrightarrow[J]{\mu_m} & & \xrightarrow[J']{\mu_m} & & \xrightarrow[J']{\mu'} \end{array} \tag{4.45}$$

It should be emphasized once more that the 1-port representation of diffusion processes like in (4.44) crucially depends on the validity of the linear approximation of J in (4.43) in terms of $\Delta\mu$. In fact, such an approximation is usually not too bad for diffusion processes, in contrast to chemical reactions where a linear expansion for example of J in (4.23) in terms of the over-all affinity $A = A^f - A^r$ would be a very poor approximation and valid only in the very vicinity of the thermodynamic equilibrium. We shall therefore represent chemical reactions involved in the following models always as 2-port elements as in (4.14). Even for diffusion processes where a linear approximation is justified, we shall frequently prefer the 2-port version associated with (4.41), the reason being simply that a consistent concentration language including only 2-port elements is more tractable than a mixed concentration and potential language where some of the 2-port elements are reduced to 1-port elements.

Problems

1) Show that two 0-junctions which are directly connected by a bond can always be combined into one single 0-junction. The same holds for two directly connected 1-junctions.
2) In contrast to an electric capacitance the material capacitance (4.5) of ideal systems is not a constant but a function of its potential or its concentration, respectively. Compare the time behaviour of the charges (concentrations) when charging a constant electric and a variable material capacitance each at constant fluxes.
3) Calculate c_1 and c_2 as a function of time for the network of (4.35) and prove that c_1 and c_2 for $t \longrightarrow \infty$ coincide with the steady state values obtained from $\bar{J} = 0$.
4) Extend the bond-graph network (4.45) of the black-box model for transport across membranes by a reaction which describes an inhibition $X_m \rightleftharpoons X_m^*$ of the transported molecule X_m in the membrane interior, X_m^* being the inhibited (or bound) molecule which is not transported. (Reformulate the diffusion processes in (4.45) as 2-ports). Can the steady state or the relaxation properties of the extended network be distinguished from that of Section 2.3?
5) The inhibition of an enzyme E by ρ molecules of a metabolic product S_n, $E + \rho S_n \rightleftharpoons E^*$, which is part of the model in Section 2.7, is a simplified version of a series of ρ reactions in which each single step adds one further molecule to the complex formed by E and S_n. Construct the bond-graph network for the detailed series of reactions and show that the results of Section 2.7 are not essentially changed if one assumes that only the free enzyme E is active.
6) Extend the bond-graph network of (4.38) to the case that the enzyme E is confined to the interior of a biological cell whereas S and P may pass the cell membrane by diffusion. Assume that all reactants are homogeneously distributed in

the intracellular medium and S and P are maintained at constant concentrations with values S_e and P_e in the extracellular medium. Calculate the steady state reaction rate for a vanishingly small value of P_e and a vanishing reverse rate of R_2. Prove that the reaction rate depends linearly on S_e at small values of S_e and saturates at high values of S_e.

7) Construct a bond-graph network for the following situation: two kinds of molecules X_1 and X_2 can diffuse across a phase boundary in such a way that for each molecule X_1 which has passed the boundary a number of n molecules X_2 has to pass the boundary in the opposite direction. Consider the case of linearized diffusion relations as in (4.43) and (4.44) and calculate the diffusion fluxes J_1 and J_2 of X_1 and X_2 as functions of the difference $\Delta\mu_1$, $\Delta\mu_2$ of the chemical potentials of X_1 and X_2 across the boundary. (Express the coupling of the fluxes by a transducer).

5. Networks for Transport Across Membranes

5.1 Pore Models

According to the generally accepted "unit membrane model" established by Danielli and Davson in 1935, a biological membrane may be considered as a double layer of lipid molecules. The characteristic feature of the lipids is their hybrid structure consisting of a polar head group to which a twofold hydrocarbon chain is attached. When forming a membrane, the lipids of the two layers are brought into contact with their hydrocarbon layers thus creating a hydrocarbon and hence hydrophobic phase in the membrane interior. The polar head groups at the exterior surfaces of the membrane are covered by layers of specific structural membrane proteins. The thickness of the whole membrane is of the order of 100 A.

The essential point for transport across such a kind of membrane is the conclusion that the hydrocarbon phase in the membrane interior represents a highly insulating barrier at least to ions and polar molecules. It is evident, therefore, that there must exist special transport facilities in the membrane for ions and polar molecules. In principle, two possibilities for such facilities can be thought of, namely, a) "carriers", i.e., shell-shaped molecules with an external hydrocarbon structure which are capable of taking up ions and molecules at the membrane surfaces and carrying them across the interior, or b) "pores", for example protein helices, which connect both sides of the membrane and through which ions and molecules can penetrate across the hydrocarbon phase.

In the present section, let us now discuss the simplest possibility of representing a pore model in terms of bond-graph network elements and ask whether the results obtained from that network essentially differ from those of the black-box model of Section 2.3. In Section 5.4 we shall then consider the simplest network realizations of carrier mechanisms. Our starting point is the assumption that a limited number of pores is present in the membrane. The pores are represented by a material capacitance with concentration X of numbers or moles of pores per membrane area. From either side of the membrane a neutral molecule with concentration c, c', respectively, can selectively jump into an empty pore thus forming a molecule-pore complex with concentration Y. Clearly, this jump process will be represented by a 2-port reaction element like that in (4.27) where the reactants 1 and 2 on the forward side will be replaced by the molecule in the adjacent solution and the empty pore, respectively, and the reactant 3 on the reverse side by the molecule-pore complex Y. Again, the number Y of complexes per membrane area is represented by a material capacitance from which the complexes can dissociate either by returning into the reverse direction of the reaction or by proceding into the same direction such that in a second reaction identical to the first one, the molecule leaves the membrane on the opposite side and an empty pore returns into its capacitance. The complete bond-graph network for this model is given as

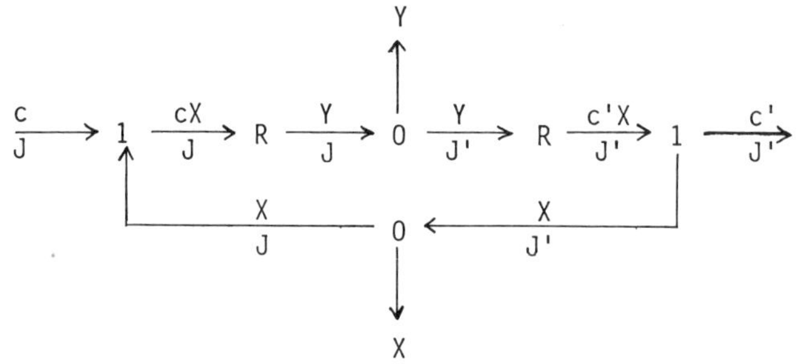

(5.1)

The reaction elements R now play a double role, namely formation and dissociation of molecule-pore complexes and diffusion across the membrane. According to the ansatz (4.26) the reaction fluxes J and J' are given as

$$J = k'cX - kY \quad , \qquad J' = kY - k'c'X \quad . \tag{5.2}$$

In (5.2) we have chosen the dissociation-formation reaction symmetrically on both sides, which need not be the case but is just the simplest possibility. For the time variation of X and Y we have according to KCL at the 0-junctions

$$\frac{dX}{dt} = - J + J' \qquad \frac{dY}{dt} = J - J' \tag{5.3}$$

from which we immediately conclude

$$X + Y = Q = \text{const} \tag{5.4}$$

where Q is the total amount of pores per membrane area.

The external capacitances c and c' for the molecule in the adjacent solutions are assumed to be infinitely large such that c, c' are constant.

Before proceeding into any detailed calculations, we immediately recognize that our model network (5.1) topologically coincides with the network (4.38) of the model for the enzyme-catalyzed reaction of Section 2.2. This means that we can formally transfer all results from Section 2.2 to the present problem and only have to interpret those results according to the physical and biological context of the pore model. In this way, we derive the steady state flux $\bar{J}$ across the pores from (2.6) by comparing (5.2) with (2.4) and translating the variables and constants of the enzyme model to that of the pore model to obtain

$$\bar{J} = \bar{J}' = \frac{k\alpha Q}{2} \frac{\Delta c}{1 + \alpha\bar{c}} \tag{5.5}$$

where $\alpha = k'/k$, $\Delta c = c - c'$, $\bar{c} = (c + c')/2$. Comparison of the steady state result (5.5) with Fick's first law, (2.16), as an ansatz for the steady state flux of the black-box model in Section 2.3 shows an additional concentration dependent term in the denominator of (5.5) due to the assumption of a limited amount of pores. For small values of the concentrations, precisely for $\alpha\bar{c} \ll 1$, (5.5) and (2.16) are effectively equivalent. At high concentrations the flux of (5.5) shows saturation similar to that in Fig. 1 of Section 2.2. From the result of the time-dependent calculations of Section 2.2 we conclude that our pore model shows exponential relaxation under nonstationary conditions with one single relaxation time τ derived from (2.15) as

$$\frac{1}{\tau} = 2k(1 + \alpha\bar{c}) \quad . \tag{5.6}$$

So far, we have always assumed that a neutral molecule penetrates the membrane across the pore. From Section 3.6 we know that in the case of an ion the chemical potential μ has to be replaced by the electrochemical potential η defined in (3.39). This means that we may immediately derive the steady state flux of an ion from (5.5) by performing the same replacement as in (2.44) to obtain

$$\bar{J} = \frac{k\alpha Q}{2} \frac{ce^{\varphi} - c'e^{-\varphi}}{1 + \frac{\alpha}{2}\left(ce^{\varphi} + c'e^{-\varphi}\right)} \tag{5.7}$$

where $\varphi = zFV/2RT$ and the potentials Φ and Φ' on the two sides of the membrane have been chosen as $\Phi = V/2$, $\Phi' = -V/2$. Eq. (5.7) tells us that for the flux J as a function of the membrane potential V we also expect a saturation at high values of V.

5.2 Pore Blocking

The network (5.1) is easily generalized to include the passage of two different molecules or ions 1 and 2 through the same pore:

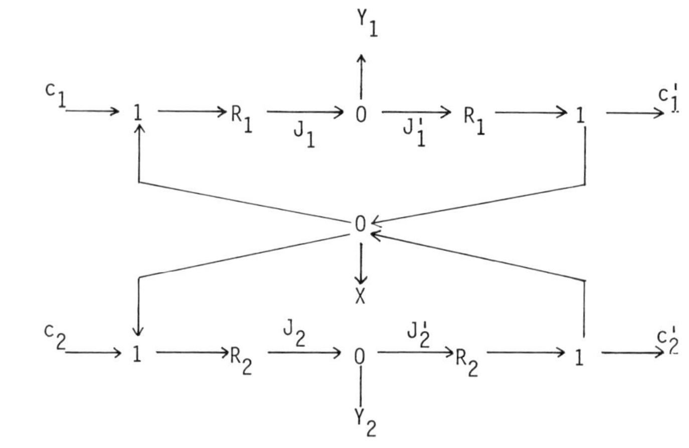

. (5.8)

For the sake of simplicity of writing we have omitted in (5.8) all concentration and flux variables which follow from KCL or KVL at the 0- or 1-junctions, respectively. Clearly, X, Y_1, Y_2 are the capacitances for empty pores and for the 1- and 2-complexes, respectively, and R_1 and R_2 are the corresponding formation and dissociation reactions with fluxes

$$\left.\begin{array}{ll} J_1 = k_1'c_1X - k_1Y_1 & J_1' = k_1Y_1 - k_1'c_1'X \\ J_2 = k_2'c_2X - k_2Y_2 & J_2' = k_2Y_2 - k_2'c_2'X \end{array}\right\} . \qquad (5.9)$$

It is clear that we now have $X + Y_1 + Y_2 = Q$. All further calculations follow directly that of Section 5.1. The steady state fluxes of molecules 1 and 2 are given as

$$\bar{J}_i = \bar{J}_i' = \frac{k_i\alpha_i Q}{2}\,\frac{\Delta c_i}{1 + \alpha_1\bar{c}_1 + \alpha\bar{c}_2} \quad , \qquad i = 1,2 \qquad (5.10)$$

where

$$\alpha_i = \frac{k_i'}{k_i} \quad , \qquad \Delta c_i = c_i - c_i' \quad , \qquad \bar{c}_i = \frac{1}{2}\,(c_i + c_i') \quad . \qquad (5.11)$$

First of all we notice that the dependence of $\bar{J}_i$ on its own concentration difference Δc_i and average concentration $\bar{c}_i$ is very much the same as in (5.5) including the phenomenon of saturation. In addition to (5.5), however, there is now an interaction between the fluxes of 1 and 2 due to the fact that both molecules compete for the same type of pores which are available only at a finite total number. Thus, increase of the average concentration $\bar{c}_1$ decreases the flux $\bar{J}_2$ and vice versa.

Within the framework of linear irreversible thermodynamics (LIT) as introduced in Section 3.9 a negative interaction would be reflected by a negative coefficient $L_{12} = L_{21} < 0$. In contrast to the scheme of LIT, however, the interaction incorporated in (5.10) is of a nonlinear type. Expansion of $\bar{J}_i$ in terms of $\Delta\mu_j = \mu_j - \mu_j'$ by making use of the ideal relationship between concentrations and potentials would lead to quadratic and higher-order interaction terms between molecules 1 and 2 since the whole expression on the right-hand side of (5.10) is at least proportional to Δc_i and thus to $\Delta\mu_i$.

Of particular interest for pore blocking in biological membranes is the case that molecules 1 and 2 are Na^+- and Ca^{2+}-ions, respectively, and that the passage of Ca^{2+}-ions through the pore is incomplete such that $J_2' = 0$ which means that the Ca^{2+} can neither leave the pore on the c_2' - side of the membrane nor enter it from there. Let us first study the consequences of $J_2' = 0$. Since now $dY_2/dt = J_2$ we have in the steady state $\bar{J}_2 = 0$ from which we immediately derive

$$\bar{Y}_2 = \alpha_2 c_2 \bar{X} \quad . \tag{5.12}$$

$\bar{X}$ is the steady state value of X. The steady state condition for Y_1, $dY_1/dt = \bar{J}_1 - \bar{J}_1' = 0$, leads to

$$\bar{Y}_1 = \alpha_1 \bar{c}_1 \bar{X} \quad . \tag{5.13}$$

Insertion of (5.12) and (5.13) into $\bar{X} + \bar{Y}_1 + \bar{Y}_2 = Q$ yields

$$\bar{X} = \frac{Q}{1 + \alpha_1\bar{c}_1 + \alpha_2\bar{c}_2} \tag{5.14}$$

which upon insertion into $\bar{J}_1$ together with (5.13) gives for the steady state flux $\bar{J}_1$ of molecule 1

$$\bar{J}_1 = \frac{k_1\alpha_1 Q}{2} \frac{\Delta c_1}{1 + \alpha_1\bar{c}_1 + \alpha_2 c_2} \quad . \tag{5.15}$$

In a second step we replace all neutral concentrations c_i and c_i' by the corresponding ionic expressions, cf. (2.44). Assuming that the empty pores are not charged and noticing $z = 2$ for Ca^{2+}, (5.12) is then replaced by

$$\bar{Y}_2 \exp\left(\frac{2F\Phi_2}{RT}\right) = \alpha_2 c_2 \bar{X} \exp\left(\frac{2F\Phi}{RT}\right) \tag{5.16}$$

where Φ_2 and Φ are the electric potentials of the Y_2 - capacitance and on the c_2-side, i.e., on the left-hand side of the membrane, respectively. From (5.16) we derive

$$\bar{Y}_2 = \alpha_2 c_2 \bar{X} \exp\left[\frac{2F(\Phi - \Phi_2)}{RT}\right] . \tag{5.17}$$

Similarly we obtain from (5.13)

$$\bar{Y}_1 = \frac{\alpha 1}{2} \bar{X} \left(c_1 \exp \frac{F\Phi}{RT} + c_1' \exp \frac{F\Phi'}{RT}\right) \exp\left(-\frac{F\Phi_1}{RT}\right) \tag{5.18}$$

where Φ' and Φ_1 are the electric potentials on the c_1' - side, i.e., on the right-hand side of the membrane and of the Y_1 - capacitance, respectively. By inserting (5.17) and (5.18) into $\bar{X} + \bar{Y}_1 + \bar{Y}_2 = Q$ and into the corresponding ionic version of the steady state flux $\bar{J} = \bar{J}_1 = \bar{J}_1'$ of (5.9),

$$\bar{J}_1 = k_1' c_1 \bar{X} \exp \frac{F\Phi}{RT} - k_1 \bar{Y}_1 \exp \frac{F\Phi'}{RT} \tag{5.19}$$

we obtain the final expression for the steady state Na^+-flux $\bar{J} = \bar{J}_1 = \bar{J}_1'$. Before evaluating this expression let us simplify the situation by choosing $\Phi = V/2$, $\Phi' = -V/2$, V being again the membrane voltage, and assuming that the potential within the membrane depends linearly on the position as in the case of a constant electric field. This is an approximation since first, the membrane could carry fixed charges and secondly, the penetrating Na^+ - and Ca^{2+}-ions could influence the electric field. Both effects would lead to a deviation from a linear dependence. With this kind of constant-field approximation we furthermore assume that the Y_1-capacitance of Na^+-pore complexes has a central position in the membrane such that $\Phi_1 = 0$ whereas the position of the Y_2-capacitance of Ca^{2+}-pore complexes remains open, its potential being written as

$$\Phi_2 = \frac{V}{2} - \frac{1^*}{1} V \tag{5.20}$$

where 1^* is the position of the Y_2-capacitance relative to the c_2-side and 1 is the thickness of the membrane. With these assumptions we now derive $\bar{J} = \bar{J}_1 = \bar{J}_1'$ from (5.19) as

$$\bar{J} = \frac{k_1 \alpha_1 Q}{2B(V)} \left(c_1 e^{\varphi} - c_1' e^{-\varphi}\right) \tag{5.21}$$

where

$$\left.\begin{aligned} B(V) &= 1 + \frac{\alpha_1}{2}\left(c_1 e^{\varphi} + c_1' e^{-\varphi}\right) + \alpha_2 c_2 e^{4l^*\varphi/l} \\ \varphi &= \frac{FV}{2RT} \end{aligned}\right\} . \tag{5.22}$$

Of particular interest is the discussion of $\bar{J}$ as a function of the membrane voltage V. Clearly, we have $\bar{J} = 0$ at the Nernst potential of Na^+,

$$V_1 = \frac{RT}{F} \ln \frac{c_1'}{c_1} \tag{5.23}$$

and a linear dependence of $\bar{J}$ on $V - V_1$ for values of V in the vicinity of V_1. For very large values of V, however, this behaviour may be drastically changed. From (5.22) we conclude that for $V \longrightarrow \infty$

$$B(V) \sim \exp\left(\frac{l^*}{l}\,\frac{2FV}{RT}\right) \tag{5.24}$$

if $l^*/l > 1/4$ and

$$B(V) \sim \exp\left(\frac{FV}{2RT}\right) \tag{5.25}$$

if $l^*/l < 1/4$ such that $\bar{J} \to 0$ for $l^*/l > 1/4$ and $\bar{J}$ saturates for $l^*/l < 1/4$ as shown in Fig. 7.

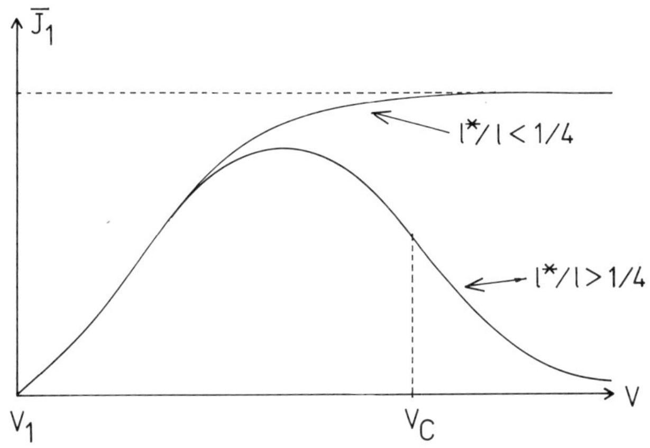

Fig. 7. Na^+-flux $\bar{J}_1$ of the pore-blocking model for two different values l^*/l of the relative Ca^{2+}-penetration depth

The critical value V_c at which for $l^*/l > 1/4$ the Na^+-flux will decrease significantly may be derived from (5.22) approximately as

$$\alpha_2 c_2 \exp\left(\frac{l^*}{l}\frac{2FV_c}{RT}\right) = \text{const} \tag{5.26}$$

such that V_c depends on the Ca^{2+}-concentration c_2 like

$$V_c \approx \frac{RT}{2F}\frac{l}{l^*}\ln c_2 + \text{const} \quad . \tag{5.27}$$

Pore blocking effects caused by Ca^{2+}-ions with a flux-voltage characteristic like that in Fig. 7 and a V_c - c_2-relation like that in (5.27) are observed in quite a lot of biological membranes. LINDEMANN, HECKMANN and SCHNAKENBERG (1972) have suggested a detailed pore model for the pore blocking in skin epithelia. The results of the detailed model are qualitatively the same as those of the simple network pore model as expressed by (5.21), (5.22) and (5.27). The adaption of the parameter l^*/l to the experimental curves gives an order of magnitude of 70 % for the relative penetration depth of Ca^{2+}.

5.3 Self-Blocking of Pores

Recently LINDEMANN (1976) has reported on a blocking phenomenon of the Na^+-pores in epithelia membranes which is not due to the action of Ca^{2+} as in the preceding section but to Na^+ itself. This conclusion was drawn from the following experimental findings: a) the flux saturates at high values of the external Na^+-concentration and b) the blocking phenomenon shows "overshooting" relaxation behaviour after a steplike increase of the external Na^+-concentration, cf. Fig. 8.

Whereas a) could be accounted for by a pore model like that in Section 5.1, an "overshooting" relaxation behaviour as indicated by the solid line in Fig. 8 cannot be brought into agreement with the relaxation behaviour of the pore network of (5.1), the reason lying in the fact that there is only one single relaxation time τ for (5.1) as shown already in Section 2.2 and as given by (5.6). Thus, the relaxation following a steplike increase of the external Na^+-concentration will always be a monotomic function of time as indicated by the dotted line in Fig. 8.

The essential point of the model which Lindemann has suggested for saturation and overshooting relaxation is the assumption that the Na^+-pores have two sites or entries for Na^+: one for transport of Na^+ across the membrane by a pore mechanism and a second one which closes the pore if occupied by a Na^+-ion. The network of this model simply reads

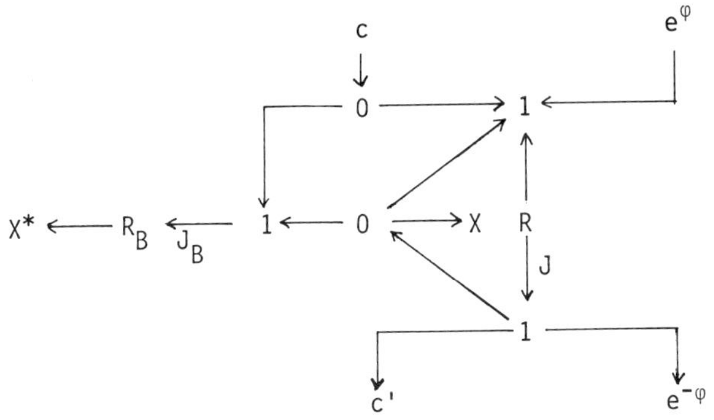

(5.28)

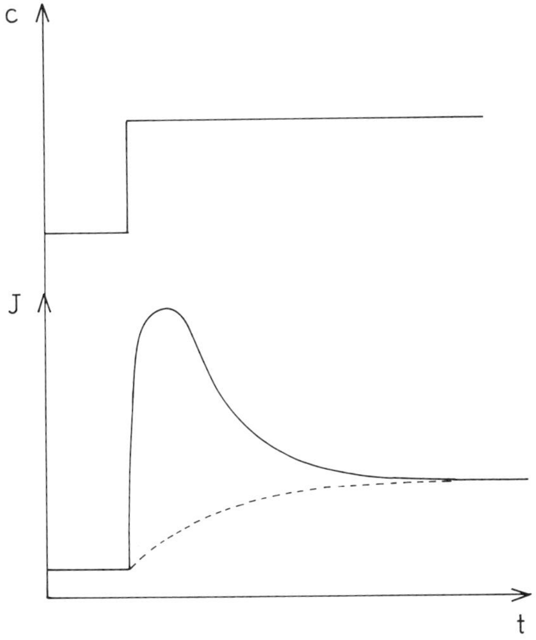

Fig. 8. Overshooting Na^+-flux J (solid line) following a steplike increase of the external Na^+-concentration c. The dotted line shows the normal relaxation behaviour of Na^+-flux in a pore model

In (5.28), the 2-port element R with flux J represents formation and dissociation of that kind of Na^+-pore complex which realized the Na^+-transport. A capacitance for the Na^+-pore complexes like Y in (5.1) has been omitted and the two reaction steps of (5.1) have been contracted into one in (5.28). Whereas in (5.1) such a simplification would deprive the model of its saturation and relaxation properties, it is uncritical in the network (5.28) as we shall see below. The flux J across R is then given by

$$J = kX\left(ce^{\varphi} - ce^{-\varphi}\right) \tag{5.29}$$

where c and c' are the external and internal Na^+-concentrations, $\varphi = FV/2RT$, V being the membrane potential, and X is the concentration of pores available for Na^+-transport per membrane area. The flux J in (5.29) satisfies the condition J = 0 at equal electrochemical potentials of Na^+ on both sides of the membrane.

In (5.28) we have chosen separate bonds for the external and internal concentrations c, c' and potentials φ, $-\varphi$. If there were no self-blocking of Na^+ we could have chosen one bond for the combined variables ce^{φ} and $c'e^{-\varphi}$. The inclusion of self-blocking on the external side, however, requires introduction of a 0-junction which divides the external Na^+-current into a transport contribution J given by (5.29) and a contribution J_B of external Na^+ with the blocking site of the pores:

$$J_B = k_B cX - k_B' X^* \tag{5.30}$$

where X^* is the capacitance of blocked pores. Note that the electrical potential does not enter into J_B which means that we assume the blocking reaction to take place on the external side on the membrane at the potential of the external solution. The kinetic equations for our model are now given as

$$\frac{dX}{dt} = -J_B \quad , \qquad \frac{dX^*}{dt} = J_B \tag{5.31}$$

from which we conclude $X + X^* = Q = \text{const}$, Q being the total number of pores per membrane area. In the steady state, we have $\bar{J}_B = 0$ and thus

$$\bar{X}^* = K_B c\bar{X} \quad , \qquad K_B = k_B/k_B' \tag{5.32}$$

which upon insertion into $\bar{X} + \bar{X}^* = Q$ yields

$$\bar{X} = \frac{Q}{1 + K_B c} \quad . \tag{5.33}$$

Insertion of (5.33) into the steady state flux $\bar{J}$ in (5.29) gives

$$\bar{J} = k\,\frac{Q}{1 + K_B c}\left(ce^{\varphi} - c'e^{-\varphi}\right) . \tag{5.34}$$

First of all, we notice that $\bar{J}$ as a function of c saturates at high values of c similar to the pore models of Sections 5.1 and 5.2. The difference between the present model and the pore models comes about when calculating the relaxation behaviour. Insertion of (5.30) and $X + X^* = Q$ into the first of (5.31) gives

$$\frac{dX}{dt} = k_B'Q - (k_B' + k_B c)X \tag{5.35}$$

which for time-independent c has the general solution

$$X(t) = \frac{Q}{1 + K_B c} + \xi \cdot e^{-t/\tau} \tag{5.36}$$

where the relaxation time τ is given as

$$\frac{1}{\tau} = k_B' + k_B c \tag{5.37}$$

and ξ is an integration constant.

Let us now apply at t = 0 a steplike increase of c to the membrane such that $c = c_1$ for $t < 0$ and $c = c_2 > c_1$ for $t > 0$. Assuming that the membrane was in a steady state at $t < 0$ we then have according to (5.36)

$$X(t) = \overline{X}_2 + (\overline{X}_1 - \overline{X}_2)e^{-t/\tau} \qquad t > 0 \tag{5.38}$$

and $X(t) = \overline{X}_1$ at $t < 0$ where

$$\overline{X}_1 = \frac{Q}{1 + K_B c_1} \quad , \quad \overline{X}_2 = \frac{Q}{1 + K_B c_2} < \overline{X}_1 \quad . \tag{5.39}$$

Inserting (5.38) into (5.29) we obtain for the time-dependent Na^+-flux

$$J(t) = \begin{cases} k\overline{X}_1\left(c_1 e^{\varphi} - c' e^{-\varphi}\right) & t < 0 \\ k\left[\overline{X}_2 + \left(\overline{X}_1 - \overline{X}_2\right)e^{-t/\tau}\right]\left(c_2 e^{\varphi} - c' e^{-\varphi}\right) & t > 0 \end{cases} \quad . \tag{5.40}$$

Plotting J(t) as a function of time we obtain an overshooting behaviour which is qualitatively similar to that in Fig. 8 except for the infinitely sharp flank at t = 0 of the calculated J(t) in constrast to the flank of finite steepness of the experimental result. This latter difference will probably be caused by some diffusion process of Na^+ on the external membrane side which could easily be included in (5.29) by an additional diffusion element.

A consistency check for the model is a plot of the measured inverse relaxation time $1/\tau$ as a function of the external Na^+-concentration c. According to (5.37) this plot should be a straight line and independent of the membrane potential. The experimental results of Lindemann are in agreement with this prediction. In particular, this confirms our previous assumption that the blocking reaction takes places at the external side of the membrane.

5.4 Carrier Models

At the beginning of Section 5.1 we have briefly discussed two alternative mechanisms for transport of a substrate, neutral molecule or ion, across a membrane, namely, pores or carriers. In both cases, the substrate is transported across the membrane as a complex either with the pore or with the carrier. What distinguishes the two mechanisms from each other is that the pores are visualized as having fixed positions with the substrate moving within them whereas the actually observed carrier molecules are known to be mobile within the membrane, either empty or loaded by a substrate molecule which can be picked up at one side of the membrane and released at the opposite side. Consequently, we design a network for a carrier mechanism by combining capacitances for empty and loaded carriers at each side of the membrane by two diffusion 2-ports, D_0 for empty and D for loaded carriers, and two identical reaction 2-ports R for the formation and dissociation of the substrate-carrier complex, one for each side of the membrane:

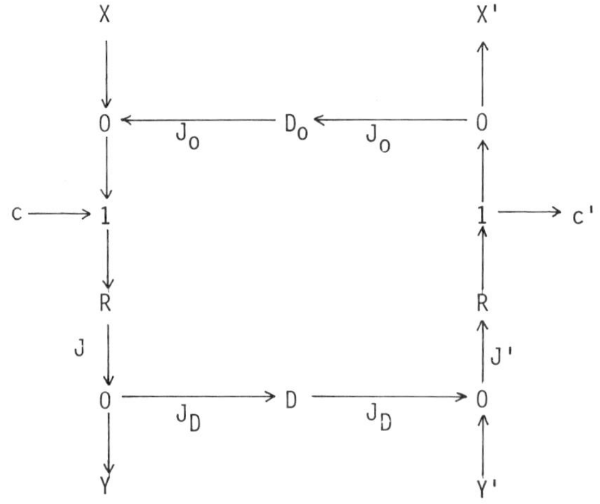

. (5.41)

X, Y, X', Y' are the capacitances of empty and loaded carriers at the two sides of the membrane, respectively, and c and c' denote the corresponding concentrations of the substrate to be transported. The constitutive relations for the fluxes of the elements D_0, D and R are given as

$$\left.\begin{array}{ll} J_0 = D_0(X' - X) & J_D = D(Y - Y') \\ J = k'cX - kY & J' = kY' - k'c'X' \end{array}\right\} \qquad (5.42)$$

where we have assumed $J_o = 0$ at $X = X'$ and $J_D = 0$ at $Y = Y'$ which means a symmetric membrane. The kinetic equations for the network (5.41) are

$$\left.\begin{aligned} \frac{dX}{dt} &= J_o - J & \frac{dX'}{dt} &= - J_o + J' \\ \frac{dY}{dt} &= - J_D + J & \frac{dY'}{dt} &= J_D - J' \end{aligned}\right\} . \tag{5.43}$$

From (5.43) we derive

$$X + X' + Y + Y' = C = \text{const} \tag{5.44}$$

which expresses the fact that the total amount C of carriers per membrane area is a finite and given constant of the model. For the steady state, we conclude from (5.43)

$$\bar{J}_o = \bar{J}_D = \bar{J} = \bar{J}' \quad . \tag{5.45}$$

Eq. (5.45) together with (5.44) are 4 linear equations for the unknown quantities $\bar{X}$, $\bar{Y}$, $\bar{X}'$, $\bar{Y}'$. The solutions of these equations is straightforward although a little tedious. The final result after having been inserted into the steady state flux reads

$$\bar{J} = \frac{D_o \alpha C}{2} \frac{\Delta c}{F(c,c')} \tag{5.46}$$

where

$$\left.\begin{aligned} F(c,c') &= 2 \frac{D_o}{k} + \frac{D_o}{D} + \left(1 + 2 \frac{D_o}{k} + \frac{D_o}{D}\right)\alpha\bar{c} + \alpha c c' \\ \Delta c &= c - c' , \quad \bar{c} = (c + c')/2 , \quad \alpha = k'/k \end{aligned}\right\} . \tag{5.47}$$

It is elucidative to compare the carrier flux (5.46) with that of the pore model in (5.5). The results are very similar in their qualitative structure; in both cases the flux $\bar{J}$ is proportional to the difference Δc of the substrate concentration and saturates for $c \to \infty$ and $c' = \text{const}$ or vice versa. The only principal difference between (5.5) and (5.46) is the fact that the denominator in (5.5) is a linear function of c and c' whereas in (5.46) it is quadratic. This means that if both c and c' tend to infinite values such that $c \sim c'$ but $c \neq c'$ the pore flux remains finite whereas the carrier flux vanishes. This latter property is due to the fact that $c \sim c' \to \infty$ blocks the cycle of the carrier action in (5.41) at the two sides of the membrane in opposite directions. Clearly, a more realistic way to distinguish a carrier from a pore mechanism would be a relaxation analysis in the case that the diffusion constants D_0 and D and the reactions constants k and k' show different orders of magnitude such that two different relaxation times can significantly be identified.

5.5 Active Transport

Very roughly speaking, active transport is a particular form of transport of some substrate uphill against increasing values of its own concentration or chemical potential. For example, quite a number of cells are able to take up K^+-ions from the extracellular surroundings although their interior K^+-concentration is much higher than that of the surroundings. This phenomenon is not just a peculiarity of biology, but of vital necessity for the cell's specific action as we have seen already for example in the case of nerve cells in Sections 2.4 and 2.5.

Uphill transport of molecules against increasing values of their concentration or potential can be discussed within the framework of linear irreversible thermodynamics (LIT) as presented in Section 3.9. If we have, for example, a 2-component system with linear phenomenological relations of the type of (4.81),

$$\left.\begin{aligned} J_1 &= L_{11}F_1 + L_{12}F_2 \\ J_2 &= L_{21}F_1 + L_{22}F_2 \end{aligned}\right\} \tag{5.48}$$

it may happen that for a sufficiently large and positive coefficient L_{12} and sufficiently large F_2 the sign of the flux J_1 is primarily determined by that of the force F_2 and possibly opposite to that of its own conjugated force F_1, i.e., $F_1J_1 < 0$. If now J_1 is the flux of some kind of molecules such that due to (3.79) F_1 has the direction of the negative gradient of the corresponding chemical potential μ_1 (at constant temperature), then J_1 is parallel to the gradient of μ_1, i.e., J_1 has a direction towards increasing values of μ_1 and its concentration c_1. On the other hand, we conclude from the second law of thermodynamics in the form of (3.78),

$$\dot{S}^{(i)} = F_1J_1 + F_2J_2 \geq 0 \tag{5.49}$$

that $F_1J_1 < 0$ is possible only at the expense of a sufficiently large and positive product $F_2J_2 > 0$. This means that the uphill transport must eventually be driven by some other downhill process. It is clear that if the system is closed with respect to these two processes, the whole process will finally come to rest as the system approaches thermodynamic equilibrium. Thus, there must be some continuously sustained external source for the uphill process in order to prevent the system from approaching equilibrium and to run the uphill process continuously. For active transport across biological membranes this source is an externally sustained non-zero value of the affinity of a chemical reaction. A simple model for such a situation is shown in the following network:

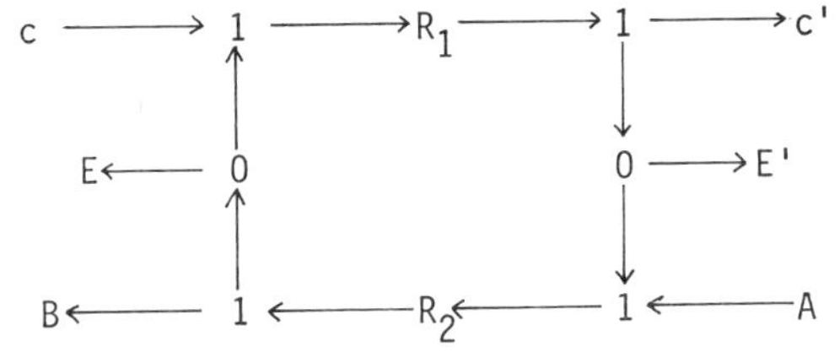

A substrate with concentrations c and c' on the two sides of the membrane is transported across the membrane by forming a complex with an activated enzyme disposed by the capacitance E. Formation, diffusion and dissociation of the complex is compressed into one 2-port element R_1. After having passed this process, the enzyme is inactivated and accumulated in a capacitance E'. A second reaction denoted as R_2 and driven by an external source $A \longrightarrow B$ re-activates the enzyme from E' to E. The reaction fluxes J_1 and J_2 are given as

$$J_1 = k_1 cE - k_1' c'E \qquad J_2 = k_2 AE' - k_2' BE \tag{5.51}$$

and the kinetic equations are

$$\frac{dE}{dt} = - J_1 + J_2 \ , \qquad \frac{dE'}{dt} = J_1 - J_2 \tag{5.52}$$

such that the total amount of enzyme is conserved:

$$E + E' = E_o = \text{const} \ . \tag{5.53}$$

In the steady state, we have $\bar{J}_1 = \bar{J}_2 \equiv \bar{J}$ which together with (5.53) is a linear system of 3 equations for the steady state values $\bar{E}$, $\bar{E}'$, $\bar{J}$. The steady state flux is immediately obtained from this system as

$$J = E_o \frac{k_1 k_2 cA - k_1' k_2' c'B}{k_1 c_1 + k_1' c' + k_2 A + k_2' B} \ . \tag{5.54}$$

Similar to the pore models, the flux is proportional to the total amount E_o of the transport facilitating enzyme. From (5.54) we also conclude that $\bar{J} > 0$ is possible even for $c < c'$ if A/B is sufficiently large such that our model actually describes active transport. Moreover, it predicts saturation of the active flux for very large values of one of the concentrations c, c', A and B.

It is elucidative to consider the linearized version of the model (5.50). To this purpose, we insert the concentration-potential relations for ideal systems into (5.51) and obtain similarly as in (2.19) or (4.43)

$$J_1 = \kappa_1\left(\exp\frac{\mu + \mu_E}{RT} - \exp\frac{\mu' + \mu_{E'}}{RT}\right) \approx \Lambda_i\left(\mu - \mu' + \mu_E - \mu_{E'}\right) \tag{5.55}$$

and similarly

$$J_2 \approx \Lambda_2\left(\mu_A + \mu_{E'} - \mu_B - \mu_E\right) \tag{5.56}$$

where μ, μ', μ_E, $\mu_{E'}$, μ_A, μ_B are the chemical potentials corresponding to the concentrations c, c', E, E', A, B and Λ_1 and Λ_2 are constant coefficients defined similarly as in (2.21) and (4.43). For the steady state, $\bar{J}_1 = \bar{J}_2$ we eliminate $\mu_E - \mu_{E'}$ from (5.55) and 5.56) to obtain

$$\bar{J} = \frac{\Lambda_1\Lambda_2}{\Lambda_1 + \Lambda_2}\left(\mu - \mu' + \mu_A - \mu_B\right) \quad . \tag{5.57}$$

Eq. (5.57) represents our first example of a linear cross-coefficient of the type L_{12}: the material flux $\bar{J}$ of the substrate is driven not only by its conjugated potential difference $\mu - \mu'$, but also by the affinity $\mu_A - \mu_B$ of a chemical reaction. In the particular case of (5.57) we have a linear relation of the type of (5.48) with $L_{12} = L_{21} = L_{11} = L_{22} > 0$.

The network (5.50) could possibly lead to the conclusion that for the performance of active transport the enzyme has to undergo a total change of position between the two sides of the membrane during its action. To avoid this misunderstanding let us decompose the total transport process of the substrate into an activated part R_1 as in (5.50) followed by a passive diffusion step denoted by a 2-port element D:

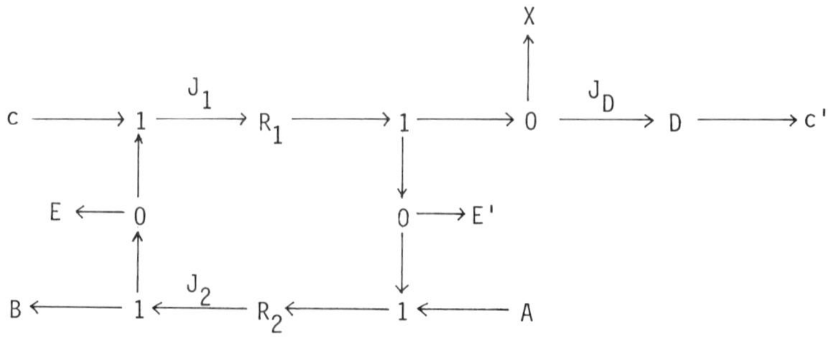

(5.58)

In (5.58), we have introduced a further capacitance X for substrate molecules somewhere within the membrane. We also see that now activated and inactivated enzyme E and E' can be present in the membrane anywhere and need not be associated with separated membrane compartments. The fluxes are now given as

$$J_1 = k_1 cE - k_1' XE' \quad , \qquad J_D = DX - D'c' \tag{5.59}$$

and J_2 as in (5.51). The kinetic equations are extended by

$$\frac{dX}{dt} = J_1 - J_D \quad . \tag{5.60}$$

The evaluation of the steady state $\bar{J}_1 = \bar{J}_2 = \bar{J}_D \equiv \bar{J}$ together with (5.53) now leads to a quadratic equation for $\bar{J}$. In order to make the result formally transparent let us choose the constants as $D' = D$ and $k_1 = k_1' = k_2 = k_2' = 1$ which is at least partially attainable by a redefinition of the external concentrations. After some transformations the final result then reads

$$\bar{J} = E_o \frac{D}{F(c,c')} (cA - c'B) \tag{5.61}$$

where

$$F(c,c') = \left[\frac{1}{4} (D(A + B) + 2D\bar{c} + B)^2 + D(cA - c'B) \right]^{1/2} + \frac{1}{2} \left[D(A + B) + 2D\bar{c} + B \right] \tag{5.62}$$

and $\bar{c} = (c + c')/2$. The result (5.61) has qualitatively the same structure as that of (5.54), especially regarding the possibility of active transport and the saturation of J. Different from (5.54) there is now an asymmetry with respect to A and B even for symmetric reactions R_2.

5.6 Hodgkin-Huxley Equations and the Coupling of Material and Electric Flux

When presenting the Hodgkin-Huxley equations of nervous excitation in Section 2.4 we noted that these equations are intended to give only a description of the excitation phenomena rather than something like a molecular explanation. In the present section, we shall translate this description into the language of bond-graphs. In doing so, we not only find a further exercise for a biological application of bond-graph networks but we will also understand that the Hodgkin-Huxley equations although not claiming to be a molecular model, can be completely expressed in the language of molecular processes like diffusion, reaction, and storage, and thus are at least very near to a molecular interpretation.

Our starting point is the three differential equations (2.39) for the probabilities p_i, i = 1,2,3, for the presence of the particles which control the conductivity of the K^+- and Na^+-channels across the membrane. Expressing this probability equivalently by the corresponding concentrations $c_i \sim p_i$ we may write

$$\frac{dc_i}{dt} = k_i\left(C_i - c_i\right) - k_i^{\prime}c_i \tag{5.63}$$

where C_i is the maximum value of the concentration of the particle of the i-type. From our considerations in Section 4.4, we immediately derive the network for (5.63) as

$$c_i^* \xrightarrow[J_i]{} R_i \xrightarrow[J_i]{} c_i \tag{5.64}$$

where

$$c_i^* = C_i - c_i \quad , \qquad J_i = k_i c_i^* - k_i^{\prime} c_i \quad . \tag{5.65}$$

From

$$\frac{dc_i}{dt} = J_i \quad , \qquad \frac{dc_i^*}{dt} = -J_i \tag{5.66}$$

we have $c_i + c_i^* = C_i$ = const and thus together with (5.65) the relation (5.63).

In the next step, we turn to the Hodgkin-Huxley relations (2.34) which together with the definitions of the conductivities g_K and g_{Na} may be written as voltage-flux relations

$$J_K = \frac{I_K}{F} = \frac{g_K^o}{F}\, c_1^4\left(V - V_K\right) \quad ; \qquad J_{Na} = \frac{I_{Na}}{F} = \frac{g_{Na}^o}{F}\, c_2^3\left(C_3 - c_3\right)\left(V - V_{Na}\right) \tag{5.67}$$

for the K^+- and Na^+-fluxes J_k, J_{Na} where V_k and V_{Na} are the K^+- and Na^+-Nernst potentials as defined in (2.33), respectively, and $V = V_a - V_e$. The factor c_1^4 in (5.67) bears some similarity to the factor X describing the concentration of open pores in (5.29) of Section 5.3 except for the fact that it is the fourth power of c_1 appearing in J_k instead of the first power of X in (5.29). On the other hand, we know from Section 4.5 how to represent stoichiometric coefficients or powers of concentrations unequal to one, namely by a transducer element. This leads us for the K^+-channel to the network

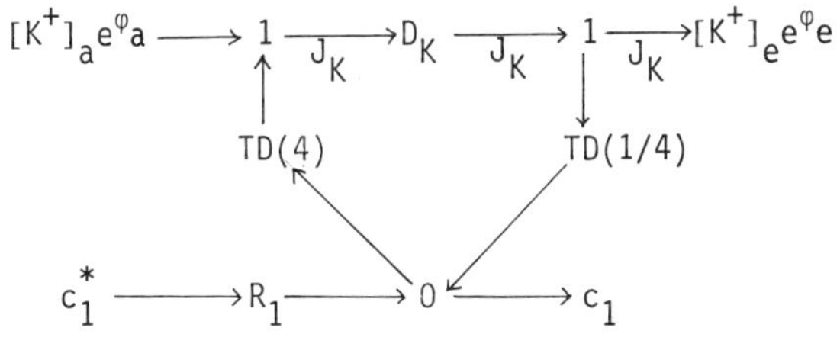

(5.68)

with

$$\varphi_{a,e} = \frac{FV_{a,e}}{RT} \tag{5.69}$$

and $[K^+]_a$, $[K^+]_e$, V_a, V_e are the values of K^+-concentration and electric potential in the axoplasm and in the external solution, respectively. The 2-port diffusion element describes the K^+-channel with a flux given as

$$J_K = D_K c_1^4 \left([K^+]_a e^{\varphi_a} - [K^+]_e e^{\varphi_e}\right) \quad . \tag{5.70}$$

Since due to KVL and KCL in (5.68) the particles with concentration c_1 are neither consumed nor produced by the diffusion process, the reaction balance of R_1 and its flux J_1 in (5.65), (5.66) remain unchanged. In order to compare (5.70) with the linear relation for J_k in (5.67), let us linearize (5.70) with respect to $V - V_K$. Making use of the definition of V_K in (2.33), choosing $V_a = V/2$, $V_e = -V/2$ and defining an average concentration of K^+ by $c_K = \left([K^+]_a [K^+]_e\right)^{1/2}$ we may write

$$\begin{aligned} &[K^+]_a e^{\varphi_a} - [K^+]_e e^{\varphi_e} \\ &= c_K \left\{ \exp\left[\frac{F(V - V_K)}{2RT}\right] - \exp\left[-\frac{F(V - V_K)}{2RT}\right]\right\} \approx c_K \frac{F}{RT} (V - V_K) \end{aligned} \tag{5.71}$$

for small $V - V_K$. Inserting (5.71) into (5.70) yields the K^+-relation of (5.67) with $g_K^o = D_K c_K F^2/RT$. It is straightforward now to represent the Na^+-channel quite analogously by

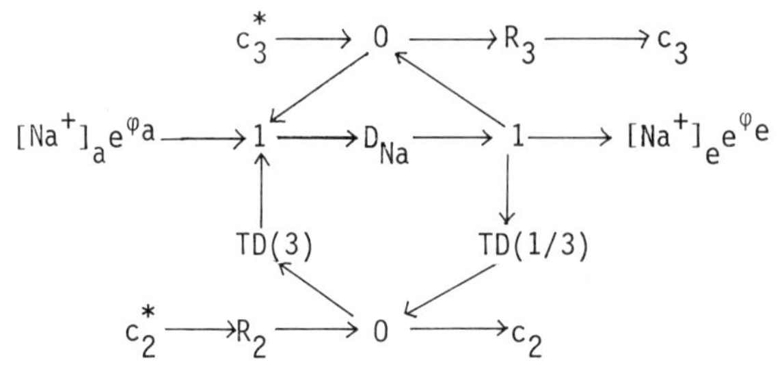

(5.72)

In a further step, we try to represent expression (2.34) for the total electric current I across the membrane by an adequate combination of the partial networks (5.68) and (5.72). Since the fluxes in (5.68) and (5.72) are of different kinds, namely K^+ and Na^+, we cannot simply add them by connecting (5.68) and (5.72) via 0-junctions. Instead, we first have to extract the electric parts of the fluxes of the partial networks (5.68) and (5.72). For the K^+-network of (5.68) this is performed as

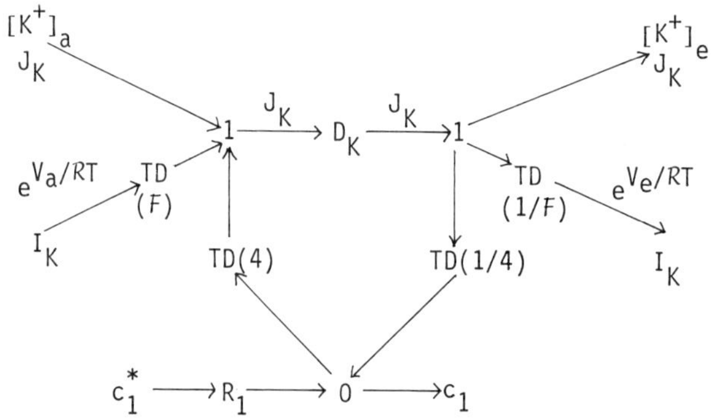

(5.73)

In the network of (5.73), I_K is the electric flux associated with the K^+-flux J_K, i.e., $I_K = FJ_K$. This relation is realized in (5.73) by the two transducers with moduli F and $1/F$, respectively, cf. (4.34) and (4.37). Regarding the concentration variables, the choice of exp $(V_{a,e}/RT)$ as the concentration equivalent of the electric ports of (5.73) directly leads back to (5.70) for the K^+-flux under the combined influence of K^+-concentrations and the electric potential.

Quite analogously we construct the electric flux I_{Na} associated with the Na^+-flux of the partial network (5.72). Summation of I_K and I_{Na} by 0-junctions leads to the total Ohmic electric flux which, in the linearized version, is given by the second and third term on the right-hand side of (2.34). What remains to be done is to include the capacitive flux $(C_e/F)dV/dt$ where C_e/F is the electric capacitance per membrane area. To this purpose let us consider the network

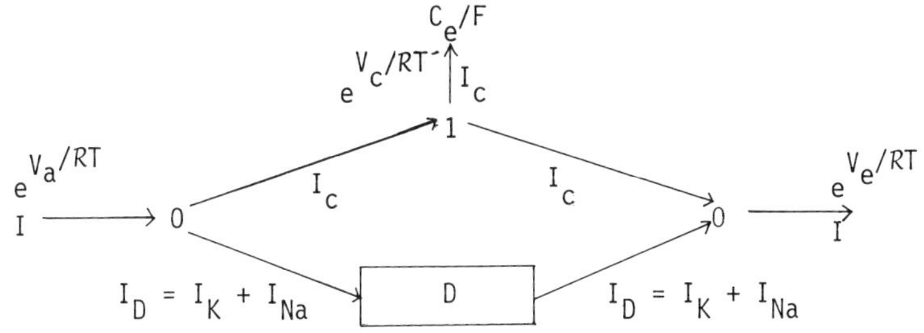

(5.74)

In (5.74), the box D represents the combination of the partial diffusion networks (5.73) and its Na^+-counterpart whereas C_e/F is an electric capacity per area with a constitutive relation

$$I_c = \frac{C_e}{F}\frac{dV_c}{dt} \quad . \tag{5.75}$$

KCL and KVL at the 0- and 1-junctions of (5.74) imply

$$I = I_c + I_D \quad , \qquad V_c = V_a - V_e = V \tag{5.76}$$

such that the total electric flux per membrane area is given by

$$I = \frac{C_e}{F}\frac{dV}{dt} + I_D \quad . \tag{5.77}$$

Replacing I_D by $F(J_K + J_{Na})$ where J_k is the linearized version of (5.70) and J_{Na} is its Na^+-counterpart, directly leads to the Hodgkin-Huxley equation (2.34). The scheme of the complete Hodgkin-Huxley network is shown in Fig. 9.

5.7 Nernst-Planck Equations

Although networks originally were intended as a language for describing processes particularly in discontinuous systems, they may also be utilized to derive equations for continuous systems in a very consistent and transparent way. As an example, let us derive the so-called Nernst-Planck equations for ionic transport in continuous media from a network construction for the underlying elementary processes. Before going into the details of the derivation we should mention that the Nernst-Planck

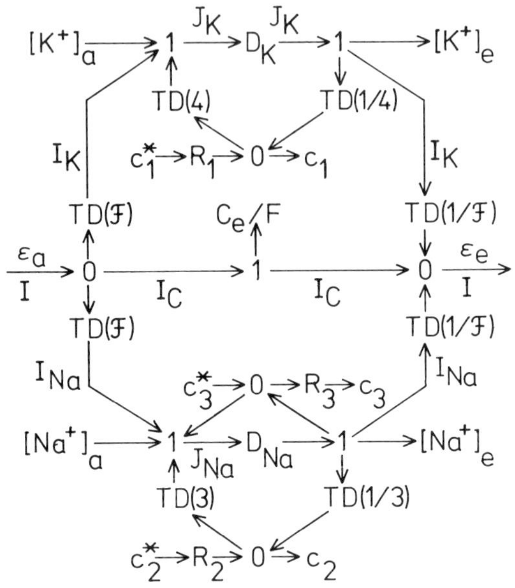

Fig. 9. Complete bond-graph network of the Hodgkin-Huxley model. The "signal flows" which cause the rate constants of R_1, R_2, R_3 to depend on the membrane voltage $V = V_a - V_e$ (cf. 2.37) are not included

equations are frequently applied to transport of ions across membranes where the membrane is not treated as a black box but assumed to have some kind of a well-defined internal structure.

Let us consider a continuous medium which is penetrated by different ions i = 1, 2, ... and which may be inhomogeneous in its composition and its transport properties. For the sake of simplicity of notation, let us assume that the allowed inhomogeneity is confined to a fixed straight direction, say parallel to the x-axis. Such situations will frequently occur in membranes with the x-axis perpendicular to the membrane plane. Let us then subdivide the medium into disk-like compartments with infinitesimal thickness dx along the x-axis and infinitely large perpendicular to the x-axis. (If the medium were arbitrarily inhomogeneous, the compartments would have to be chosen as infinitesimal rectangular volume elements dx dy dz). Each compartment is considered now as a combination of an infinitesimal resistance and an infinitesimal capacitance element for each kind of the penetrating ions. Whereas the resistance of the medium is assumed to be a joint property of the material and electric interactions of the ion with the medium, separate capacitances are introduced for the storage of matter and electric charge: one material capacitance for each kind of ions i = 1,2, ... and a common electric capacitance for the net amount of electric charge. If only one kind of ion with a valency z_i were present, the network for a single compartment would then be given by

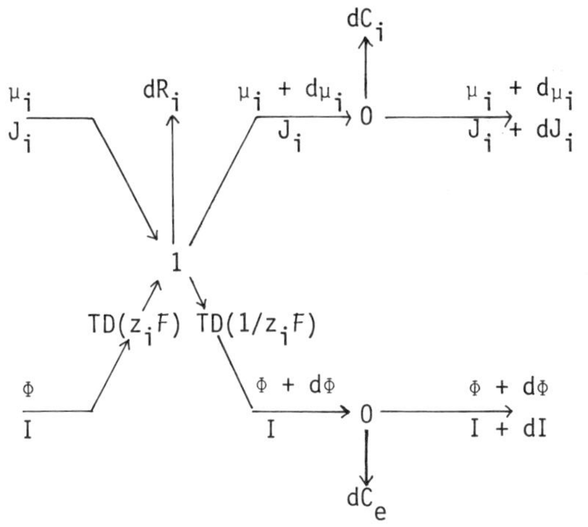

. (5.78)

In (5.78), μ_i and Φ are the chemical and the electric potentials, J_i and I are the material and the electric fluxes, respectively. The potentials and fluxes change their values by infinitesimal increments $d\mu_i$, $d\Phi$, dJ_i, dI along the compartment. The potential-flux language is chosen instead of the concentration-flux language since the Nernst-Planck equations represent a linear theory in the sense of 1-port diffusion elements, cf. Section 4.6. The values of potentials and fluxes at the bonds of (5.78) have been chosen such that KCL and KVL at the 0- and 1-junctions are automatically satisfied. Moreover, use has been made in (5.78) of what we have learned already in Section 5.6 about the coupling of material and electric fluxes, cf. (5.73).

The infinitesimal material and electric capacitances dC_i and dC_e respectively, are written as

$$dC_i = \gamma_i F \, dx \quad , \qquad dC_e = \gamma_e F \, dx \tag{5.79}$$

where γ_i and γ_e are the corresponding capacitances per volume and F is the area of the compartment perpendicular to the x-axis. Similarly, we write

$$dR_i = \rho_i dx/F \tag{5.80}$$

for the infinitesimal resistance where ρ_i is the specific resistivity.

The basic equation describing our theory will now be obtained from the constitutive equations of the elements dC_i, dC_e, dR_i. The constitutive equation of the material capacitance dC_i is obtained as

$$-dJ_i = dC_i \cdot \frac{\partial}{\partial t} (\mu_i + d\mu_i) \tag{5.81}$$

which upon cancelling second-order differentials, inserting dC_i from (5.79) and expressing dJ_i as $dJ_i = (\partial J_i/\partial x)dx$ yields

$$\gamma_i \frac{\partial \mu_i}{\partial t} + \frac{\partial j_i}{\partial x} = 0 \tag{5.82}$$

where $j_i = J_i/F$ is the material flux density. In (5.81) and (5.82) we have written partial derivatives $\partial/\partial t$ and $\partial/\partial x$ since we now have a twofold dependence of all quantities on position x and time t.

From (5.81) we also conclude that

$$\mu_i dC_i = \gamma_i \mu_i F \, dx \tag{5.83}$$

is the number of moles of ions of kind i contained in the compartment. This implies that $c_i = \gamma_i \mu_i$ is the molar concentration of the ion at position x and at the time t. Thus, (5.82) can be written as

$$\frac{\partial c_i}{\partial t} + \frac{\partial j_i}{\partial x} = 0 \tag{5.84}$$

which is nothing else than the continuity equation expressing the conservation of material of the ions of kind i. In the same way we derive the continuity equation for the electric charge from the constitutive equation of the electric capacitance dC_e as

$$\gamma_e \frac{\partial \Phi}{\partial t} + \frac{\partial j}{\partial x} = 0 \tag{5.85}$$

where $j = I/F$ is the electric flux density and $\rho = \gamma_e \Phi$ is the volume density of the electric charge of the penetrating ions.

Next we formulate the constitutive equation for the resistance element dR_i. The potential drop $dR_i \cdot J_i$ of the bond connecting dR_i with the 1-junction follows from KVL as

$$dR_i \cdot J_i = -d\mu_i - z_i F d\Phi \quad . \tag{5.86}$$

Upon writting $d\mu_i = (\partial \mu_i/\partial x)dx$, $d\Phi = (\partial \Phi/\partial x)dx$ and inserting (5.80) we obtain from (5.86)

$$j_i = -\frac{1}{\rho_i}\left(\frac{\partial \mu_i}{\partial x} + z_i F \frac{\partial \Phi}{\partial x}\right) = -\frac{1}{\rho_i}\frac{\partial \eta_i}{\partial x} \tag{5.87}$$

where $\eta_i = \mu_i + z_i F\Phi$ is the electrochemical potential of the ion of kind i, cf. (3.39). Eq. (5.87) may be considered a generalized Ohm's law: the local flux density is porportional to the negative gradient of the electrochemical potential, the proportionality factor being the specific conductivity $\sigma_i = 1/\rho_i$. The usual form of the Nernst-Planck equations is obtained from (5.87) by inserting the ideal relationship (3.77) between the chemical potential μ_i and the concentration c_i such that

$$\frac{\partial \mu_i}{\partial x} = \frac{RT}{c_i} \frac{\partial c_i}{\partial x} \tag{5.88}$$

provided that the reference potential μ_i^o is independent of the position x. Eq. (5.87) together with (5.88) gives

$$j_i = -D_i \left(\frac{\partial c_i}{\partial x} + \frac{z_i F c_i}{RT} \frac{\partial \Phi}{\partial x} \right) \tag{5.89}$$

which is the usual form of the Nernst-Planck equations with the diffusion coefficient D_i given as

$$D_i = \frac{RT}{\rho_i c_i} \quad . \tag{5.90}$$

Comparison with (2.21) shows that (5.90) is the continuous generalization of the discrete model relation (2.21).

From the constitutive relation of the transducers in (5.78) we derive $I = z_i F J_i$ or, if several kinds of ions are penetrating through the medium,

$$I = \sum_i z_i F J_i \tag{5.91}$$

which also implies

$$\rho = \sum_i z_i F c_i \tag{5.92}$$

for the electric charge density of the ions. For most of the materials which are permeable for ions, it is physically reasonable to require electroneutrality, i.e. vanishing electric net charge densities $\rho = 0$. (The condition of electroneutrality has to be expressed as $\rho + \rho_o(X) = 0$ if there is a density $\rho_o(X)$ of fixed charges present in the material). Because of $\rho = \gamma_e \Phi$, cf. (5.85), $\rho = 0$ leads to

$$\frac{\partial j}{\partial x} = 0 \qquad (5.93)$$

which allows for a finite but position-independent electric flux density $j = I/F$.

For further reading, particularly concerning solutions of the Nernst-Planck equation under rather general physical conditions in membrane systems, the very detailed monograph of R. SCHLÖGL (1964) is recommended.

Problems

1) Determine the relaxation behaviour of the networks for the pore and carrier models and for active transport. Prove that under any external conditions the state of the network exponentially tends into the steady state by solving the full equations of motion or that for the differential fluctuations as in Sections 2.6 and 2.7.
2) Extend the network (5.28) for the self-blocking of pores by decomposing the reaction-diffusion process R into two partial steps and introducing an additional capacitance Y for the complexes in between them. Show that this extension leaves the results of Section 5.3 qualitatively unchanged.
3) Construct a network for different substrates S_1 and S_2 competing for the same carrier. Show that the interaction between the steady state fluxes $\bar{J}_1$ and $\bar{J}_2$ of S_1 and S_2 is nonlinear as in the case of the pore model in Section 5.2.
4) Derive the results (5.61) and (5.62) from the network in (5.58).
5) Construct an extended network from (5.50) in which an additional reaction R_3 describes a direct decay of the activated enzyme E into the inactivated form E' without having transported a substrate. Calculate the linear coefficients for this case in analogy to (5.55) and (5.56).
6) Construct a network which includes the effect of the membrane voltage V on the reactions R_1, R_2, R_3 of the controlling particles in the Hodgkin-Huxley theory by assuming that the controlling particles carry an electric charge and that the reactions R_1, R_2, R_3 are associated with a change in position. (Compare problem 3 of Chapter 2).
7) Solve the Nernst-Planck equations (5.89) for the case of a given constant electric field $E = -\partial\Phi(x)\partial x$ = const, but position-dependent concentration $c_i(x)$.
8) Let two kinds of neutral molecules A and B penetrate through a material which may be inhomogeneous along the x-direction. Assume that a chemical reaction $A \rightleftharpoons B$ may transfer A into B and vice versa at every position x of the material. Construct a network for the continuous system involving now storage and diffusion of both A and B. Derive the partial differential equations for the concentrations $c_A(x,t)$ and $c_B(x,t)$ of A and B and find simple stationary solutions.

6. Feedback Networks

6.1 Autocatalytic Feedback Loops

The qualitative global behaviour of the networks which we have developed in Sections 5.1 to 5.5 for pore and carrier mechanisms and active transport in membranes is characterized by the existence of a uniquely determined steady state into which the state of the system will tend for $t \longrightarrow \infty$ irrespective of its initial state. We have explicitly shown this behaviour only for the case of the self-blocking of pores in Section 5.3 but the analogous calculations may readily be performed in the same way for the other cases mentioned above, cf. problem 1 of Chapter 5.

A steady state into which all initial states of the system will evolve for $t \longrightarrow \infty$ is called globally and asymptotically stable. Such a situation is a particularly simple form of the stability behaviour of a dynamical system. It includes cases where the steady state is not unique; but there are multiple steady states among which, however, only one is globally and asymptotically stable. The other steady states are then instable: any arbitrarily small fluctuation drives the system out of them into the stable state.

In Section 2.5 we have studied a model for nervous excitation which shows multiple steady states under certain external conditions (cf. Fig. 6) the multiplicity being 3 in this particular case. In the context of Fig. 6, we also mentioned that by a relaxation analysis the intermediate of the three steady states can be shown to be unstable. The remaining two steady states are stable but not globally stable since the very definition of global asymptotical stability excludes the possibility of two simultaneous globally and asymptotically stable states. The relaxation analysis shows that in this situation of two stable steady states, roughly speaking, the system approaches that one to which its initial state is nearest. This type of stability of a steady state is called local asymptotical stability. By the term "local" one states that in contrast to global stability the system approaches a steady state only if it starts somewhere in a more or less restricted vicinity of the steady state. The term "asymptotical" indicates that the approach under this condition is complete for $t \longrightarrow \infty$.

The appearance of multiple locally and asymptotically stable steady states is essential for the excitation model of Section 2.5 since these two steady states represent the ground and the excited state of the nervous membrane. The same type of stability behaviour is found for the Hodgkin-Huxley model in Sections 2.4 or 5.6. Thus, from the point of view of biological application, this type of stability has to be distinguised fundamentally from that with one globally and asymptotically stable steady state.

When discussing the model for the control of metabolic reaction chains in Section 2.7 we have become acquainted with a third type of stability of a steady state. In this model, there always exists a unique steady state, cf. (2.62), (2.63). For concentrations S of the external substrate below a critical value S_c, i.e. $S < S_c$, the steady state is globally and asymptotically stable, whereas for $S > S_c$ the state of the system leaves the steady state in a spiral motion in the phase space approaching an undamped periodically oscillating trajectory, a so-called limit cycle, cf. (2.75), (2.76).

The considerations throughout this chapter will be devoted to networks which produce either multiple steady states or limit cycles. The leading idea for these considerations will be the experience that such a stability behaviour or, from the point of view of systems with one globally and asymptotically stable steady state, such an instability behaviour is caused by a feedback coupling within the network. In the language of network topology, feedback manifests itself as a closed loop of bonds, junctions and elements. The reverse conclusion, however, does not hold; the networks of the pore and carrier models and that of active transport in Sections 5.1 to 5.5 actually involve closed loops, but nevertheless show a unique globally and asymptotically stable steady state. The simplest case of a nontrivial feedback loop, i.e., a feedback loop which really leads to multiple steady states or limit cycles, is an autocatalytic reaction

$$A + \nu X \rightleftharpoons (\nu + 1)X \quad . \tag{6.1}$$

Eq. (6.1) means that for $\nu \geq 1$ species X catalyses its own production from some other species A. If, however, (6.1) is the only reaction of the system and A is externally supplied at a constant concentration, the steady state of (6.1) would be an equilibrium state since the steady state flux of (6.1) would vanish. In order to have a nonequilibrium steady state we add a second reaction to (6.1),

$$X \rightleftharpoons B \tag{6.2}$$

which describes a simple decay of the species X and where B is again externally supplied at a constant concentration. According to the network rules developed in Chapter 4, the network for the combination of (6.1) and (6.2) is given as

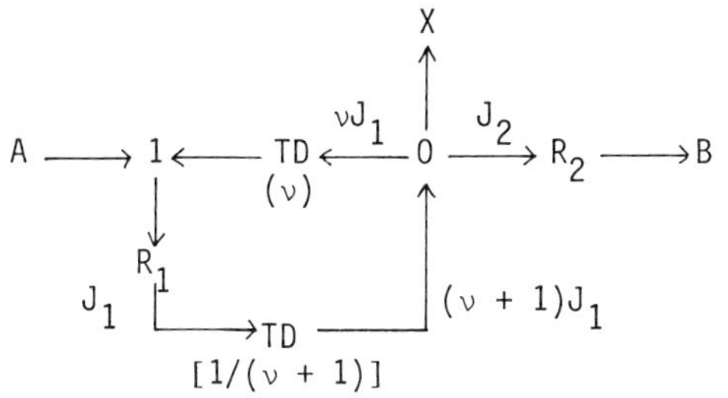

(6.3)

where R_1 and R_2 are the reaction 2-port of the reaction (6.1) and (6.2). Making again use of the simple concentration product ansatz, the reaction fluxes J_1 and J_2 of R_1 and R_2 are given by

$$J_1 = k_1AX^{\nu} - k_1'X^{\nu+1} \quad , \qquad J_2 = k_2X - k_2'B \tag{6.4}$$

where X denotes the concentration of the internal species X. The equation of motion for X, i.e., its total time derivative, is obtained from (6.4) and the network (6.3) as

$$\frac{dX}{dt} = \Phi(X) \equiv J_1 - J_2 = k_1AX^{\nu} - k_1'X^{\nu+1} - k_2X + k_2'B \quad . \tag{6.5}$$

From (6.5) we conclude $\Phi(0) = k_2'B > 0$ and $\Phi(X) \to -\infty$ as $X \to +\infty$. This means that there exists at least one zero of $\Phi(X)$, $\Phi(\overline{X}) = 0$, i.e., one steady state $\overline{X}$ in $X > 0$.

For $\nu = 1$ which is the simplest case of an autocatalytic reaction in (6.1), $\Phi(X)$ becomes a polynomial of second order with two zeros $\Phi(\overline{X}_{1,2}) = 0$ given as

$$\overline{X}_{1,2} = \frac{k_1A - k_2}{2k_1'} \pm \left[\left(\frac{k_1A - k_2}{2k_1'}\right)^2 + \frac{k_2'B}{k_1'}\right]^{1/2} \quad . \tag{6.6}$$

From (6.6) we immediately conclude that under any external conditions for A and B $\overline{X}_{1,2}$ are real and $\overline{X}_1 > 0$, $\overline{X}_2 < 0$. This means that only $\overline{X}_1$ is a physically meaningful steady state. For $0 < X < \overline{X}_1$ we have $\dot{X} = \Phi(X) > 0$ and for $\overline{X}_1 < X$ $\dot{X} = \Phi(X) < 0$ such that the state of the system in $X > 0$ always approaches $\overline{X}_1$ for $t \longrightarrow \infty$, cf. Fig. 10 (a). Despite of the autocatalytic feedback loop which is present in (6.1) even for $\nu = 1$ we thus have a unique steady state in $X > 0$ which is globally and asymptotically stable.

The situation may become quite different for $\nu = 2$ where $\Phi(X)$ is a third-order polynomial. We know already that there exists at least one zero of $\Phi(X)$ in any case, but for appropriate choices of the constants in (6.5), it may happen that

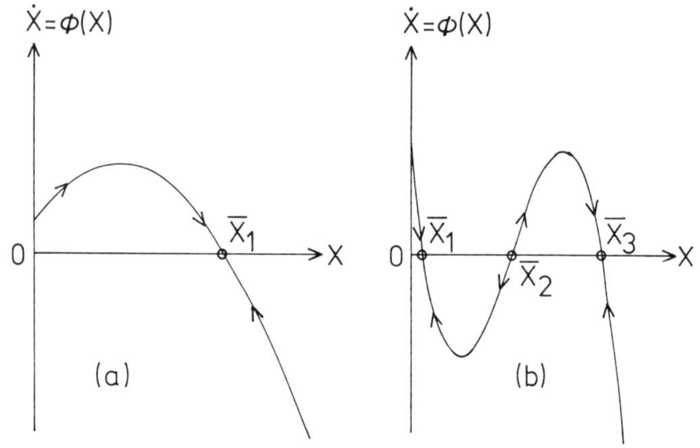

Fig. 10. Time changes $\dot{X} = \Phi(X)$ of a concentration X involved in an autocatalytic loop as a function of X, (a) for $\nu = 1$, (b) for $\nu = 2$. The arrows at the curves indicate the corresponding direction of evolution

$\Phi(X)$ for $\nu = 2$ shows three zeros in the region $X > 0$. A necessary prerequisite for such a situation is the existence of an interval in $X > 0$ where $\Phi'(X) \equiv d\Phi(X)/dX > 0$ which is guaranteed if $A^2 > 3k_1'k_2/k_1^2$ as the reader may easily verify. Although this condition is only necessary but not sufficient, it shows the tendency that the appearance of three steady states becomes increasingly probable with increasing values of A, i.e., with increasing distance from an equilibrium situation. Fig. 10 (b) shows a case with three steady states $\overline{X}_1 < \overline{X}_2 < \overline{X}_3$ for $\nu = 2$. Since we have $\Phi(0) > 0$ and $\Phi(X) \to -\infty$ for $X \to +\infty$ it follows that $\Phi(X) > 0$ in $0 < X < \overline{X}_1$ and $\overline{X}_2 < X < \overline{X}_3$ whereas $\Phi(X) < 0$ in $\overline{X}_1 < X < \overline{X}_2$ and $\overline{X}_2 < X$. This means that $\overline{X}_1$ and $\overline{X}_3$ are locally and asymptotically stable whereas $\overline{X}_2$ is instable. From Fig. 10 (b) we also immediately derive the local stability region of $\overline{X}_1$ and $\overline{X}_3$, namely $0 \le X \le \overline{X}_2$ for $\overline{X}_1$ and $\overline{X}_2 < X$ for $\overline{X}_3$.

The appearance of two stable steady states $\overline{X}_1$, $\overline{X}_3$ allows the system to exist in two phases with different densities $\overline{X}_1$ and $\overline{X}_3$ of the species X. It may even happen that these two phases coexist in the same system separated by a phase boundary. The whole situation is very similar to the phenomenon of phase transitions in equilibrium systems such as gas-liquid or liquid-solid systems. According to this similarity, the phenomenon of different phases in a nonequilibrium system is called a nonequilibrium phase transition or a "dissipative structure". Clearly, the inclusion of coexistence between $\overline{X}_1$ and $\overline{X}_3$ and of phase boundaries into our theory requires the introduction of additional diffusion terms into the equation of motion (6.5) in order to account for spatial variations of X. The analogies between our autocatalytic system (for $\nu = 2$) and equilibrium phase transitions have been worked out by F. SCHLÖGL (1972) on a phenomenological and by JANSSEN (1974) on a stochastic level.

Other choices of ν with $\nu \ge 3$ in (6.3) or (6.5) do not lead to qualitatively different types of autocatalytic systems than those for $\nu = 1$ and $\nu = 2$, not even independent choices of the moduli of the two transducers in (6.3). We therefore restrict ourselves to biological applications of the cases $\nu = 1$ and $\nu = 2$ which we shall call in the following A1 and A2 elements. As regarding the biological application of A1 we know already from Section 2.6 that A1 without the R_2-reaction may serve as a reasonable ecological simulation for a reproducing population which shows saturation. A direct application of A2 will be discussed in the following section. In the remaining parts of this chapter, we shall treat the A1 and A2 elements as "integrated circuits" within " super networks" and thereby generate models of biological interest from a network topological starting point.

6.2 An Autocatalytic Excitation Model

The steady state behaviour of the autocatalytic element A2 introduced in the preceding section is very similar to that of the Ising model disccused in Section 2.5: in both systems the parameters can be chosen such that three steady states appear among which the intermediate one is instable, but the remaining two are at least locally and asymptotically stable. In Section 2.5, these two stable steady states served as a basis for an excitation model such that one of them represents the ground state of active centers in the membrane and the other one the excited state. It is very suggestive now to think of an excitation model on the basis of the A2 element instead of the Ising model. Let us construct such a model and let us choose the external interactions of this model in the same way as in the case of the Ising model in Section 2.5, i.e., let us assume the excitation transition to be coupled to an exchange of a Ca^{2+} by two K^+ ions per active center of the nerve membrane. The bond-graph network of this model is given as

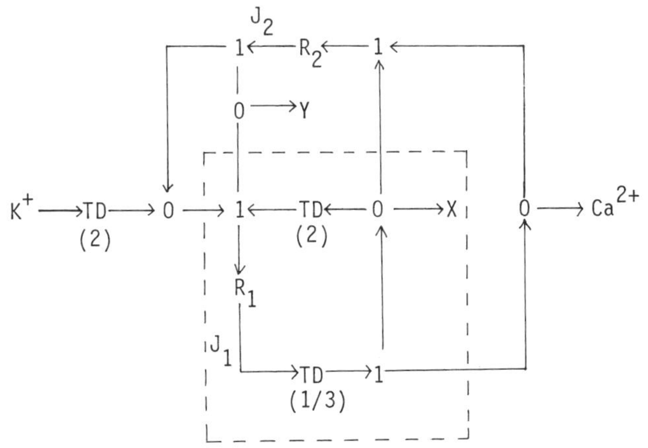

. (6.7)

In (6.7) we have enclosed the core of the model, the autocatalytic excitation transition of the A2 type, by a dotted line. The capacitances X and Y are the pools for excited and nonexcited active centers, and R_2 is a decay process of excited centers to non-excited ones. According to the model construction of the Ising type in Section 2.5, excitation via R_1 is coupled with an uptake of two K^+ and a release of a Ca^{2+} and vice versa for the decay via R_2. It will not affect our model to allow for the reverse reactions along R_1 and R_2 and we shall actually include them in the following analysis, although a reasonable physiological interpretation of the processes will lead to extremely small reverse rate constants. From (6.7) we derive

$$\frac{dX}{dt} = J_1 - J_2 \quad , \qquad \frac{dY}{dt} = - J_1 + J_2 \tag{6.8}$$

such that the total amount of active centers, X + Y = A, remains unchanged. For J_1, J_2 we make the standard product ansatz:

$$J_1 = k_1 X^2 Y - k_1' X^3 \quad , \qquad J_2 = k_2 X - k_2' Y \tag{6.9}$$

where in accordance with (2.42) the k_i, k_i' for i = 1,2 are functions of the ionic concentrations and the membrane voltage

$$\left.\begin{aligned} k_1 &= \alpha_1 c_{1a}^2 e^{2\varphi} + \beta_1 c_{1a} c_{1e} + \gamma_1 c_{1e}^2 e^{-2\varphi} \\ k_2' &= \alpha_2 c_{1a}^2 e^{2\varphi} + \beta_2 c_{1a} c_{1e} + \gamma_2 c_{1e}^2 e^{-2\varphi} \\ k_1' &= \lambda_1 c_{2e} e^{-2\varphi} \quad , \qquad k_2 = \lambda_2 c_{2e} e^{-2\varphi} \end{aligned}\right\} \quad . \tag{6.10}$$

In (6.10), we have set $c_{2a} = 0$: almost vanishing concentration of Ca^{2+} in the axoplasm. For further definitions of the symbols in (6.10), cf. (2.39).

The steady state condition for the model is obtained from (6.8) as $\bar{J}_1 = \bar{J}_2$ which upon inserting expressions (6.9) and $\bar{X} + \bar{Y} = A$ yields

$$\Phi(\bar{X}) \equiv - (k_1 + k_1')\bar{X}^3 + k_1 A \bar{X}^2 - (k_2 + k_2')\bar{X} + k_2' A = 0 \quad . \tag{6.11}$$

In order to get a qualitative impression of what happens in the model let us discuss the steady state value $\bar{X}$ as a function of the membrane voltage V which is involved in the rate constants of (6.10) by $\varphi = FV/2RT$, cf. (2.40). For large values of V, i.e., increasing potential of the axoplasm, we have $k_1' \approx k_2 \approx 0$ and (6.11) is reduced to

$$\Phi(\overline{X}) \approx (k_1\overline{X}^2 + k_2')\ (A - \overline{X}) = 0 \tag{6.12}$$

with the solution $\overline{X} \approx A$, i.e., almost all active centers are excited. For very large negative values of V, all rate constants in (6.10) remain finite. Assuming that $\gamma_1 c_{1e}^2$ and $\gamma_2 c_{2e}^2$ are small compared to $\lambda_1 c_{2e}$ and $\lambda_2 c_{2e}$ we may neglect then k_1 and k_2' which reduces (6.11) to

$$\Phi(\overline{X}) \approx -\ (k_1'\overline{X}^2 + k_2)\overline{X} = 0 \tag{6.13}$$

and hence $\overline{X} \approx 0$; almost all active centers are in the ground state. According to the general discussion of the A2-element in Section 6.1 we expect an intermediate voltage region with a triple-valued steady state $\overline{X}$ in two stable branches near $\overline{X} \approx 0$ and $\overline{X} \approx A$ and an instable central branch. This qualitative discussion is confirmed by a numerical evaluation of (6.11) shown in Fig. 11 for a particular choice of the parameters.

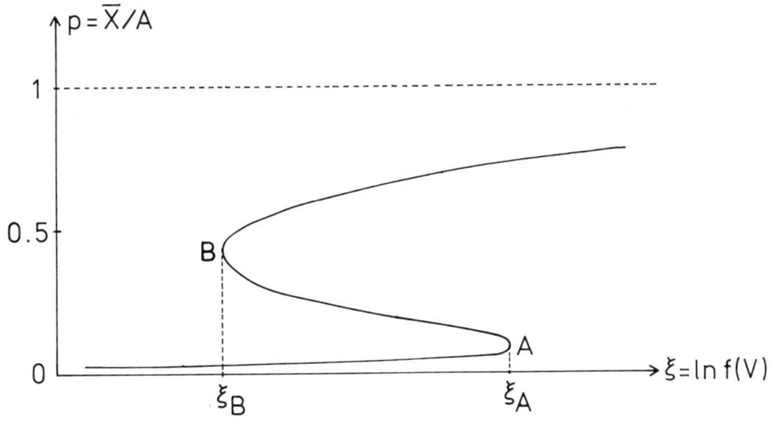

Fig. 11. Fraction $p = \overline{X}/A$ of excited active centers in the A2-model of membrane excitation. The variable $\xi = \ln f(V)$ is defined as in Section 2.5, (2.43), cf. Fig. 6

Although the $\overline{X}/A$ - V diagram of Fig. 11 is less symmetric compared to that of Adam's Ising model, cf. Fig. 6, BEBBER (1975) has shown that the parameters of the A2-model can be chosen in such a way that its time-dependent excitation curve, i.e., the solution of (6.8), becomes almost indistinguishable from that of the Ising model.

Having presented the A2-model of excitation by a bond-graph network in (6.7) it is suggestive to ask for an explicit bond-graph representation of the Ising model. The only possibility for such a representation is the almost trivial scheme

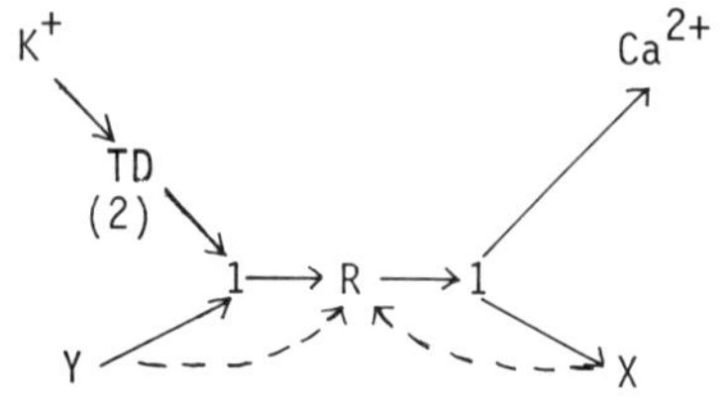

(6.14)

where the dotted lines indicate a parametric coupling of the rate constants of the reactions R to the pools of X and Y as expressed by (2.42) with p = X/A and 1 - p = Y/A. The reason why an explicit representation of this parametric coupling in terms of our bond-graph elements capacitance and reaction 2-ports is impossible lies in the fact that the Ising model cooperative coupling mechanism is an instantaneous, nondissipative energetic coupling as to be distinguished from the dissipative cooperative autocatalytic mechanism in the A2-element.

A serious biochemical objection to be raised against both the Ising and the A2-model says that a direct positive feedback coupling as assumed in both models will hardly be realized in biology. Instead, the relatively small number of known effective autocatalytic reactions should be considered as over-all balances of a complex combination of many partial reactions wherein the most frequent form of interaction between different kinds of molecules is that of inhibition. Now from the network point of view it is evident that an effective positive feedback can be generated by twice a negative inhibition interaction put in series. Actually, such a realization of an excitation model has been proposed by McILROY (1970 a, b). HOOK (1975) has reduced this model to its skeleton bond-graph structure by omitting all details of the model which are not essential for its over-all cooperativity. Quite a similar reduction procedure has been applied to a model suggested by LAVENDA (1972) for the control of biosynthesis of proteins. The comparison shows that both McIlroy's and Lavenda's models are based upon the same idea of cooperativity via twice a negative interaction. Thus, bond-graph networks are not only a method for constructing simple models but may also be expected to be helpful when analyzing and classifying existing models with respect to their essential mechanisms.

6.3 Networks of A1-Elements

In this section we shall construct networks involving A1-elements as "integrated circuits". Let us stress once more that the A1-element placed between to external reservoirs A and B is globally and asymptotically stable, cf. Fig. 10.

The simplest way to form a network of A1-elements is to put two A1-elements in series:

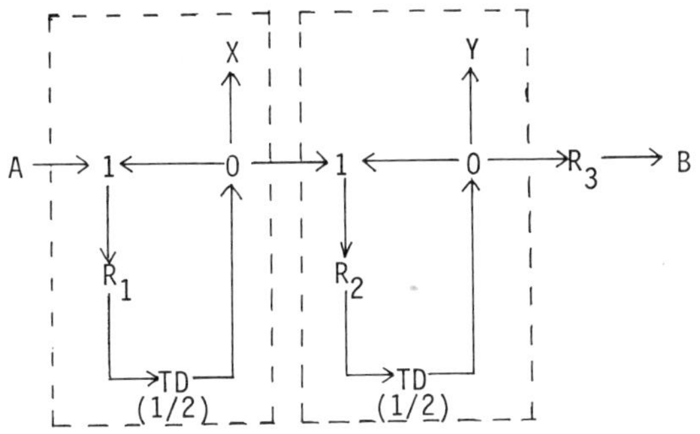

. (6.15)

Again we have enclosed the two A1-elements with their variables X and Y by dotted lines. Making the usual product ansatz for the fluxes J_1, J_2, J_3 along R_1, R_2, R_3 we immediately recover the flux relations (2.53) of the Volterra-Lotka model including the reverse rates or saturation terms. Clearly, this construction principle can be extended to generate Volterra-Lotka chains with an arbitrary number of species. It creates no difficulties to include the case of branching chains such that at a branching point two or more predator species are competing for the same prey or one single predator has the choice between two or more preys.

Concerning the stability behaviour of extended Volterra-Lotka chains we have mentioned already in Section 2.6 that the inclusion of reverse rates or saturation terms tends to stabilize the chain such that the periodic oscillations in the absence of reverse rates become damped. Clearly, the construction of Volterra-Lotka chains on the basis of A1-elements as integrated circuits automatically introduces reverse rates. As LEISEIFER (1975) has shown it is not necessary to have the reverse rates in each of the A1-elements of the network. Introducing only one single reverse rate in one of the reactions which connects an internal variable with an external source or sink guarantees the existence of a unique, globally and asymptotically stable steady state provided that the super-network of A1-elements does not generate new loops of A1-elements. From a network point of view, this latter finding could be interpreted by saying that stability is a property which can be transmitted in networks from one element to the others.

On the other hand, we now conclude that for producing an instability we necessarily have to generate a closed loop in the super-network of A1-elements. Again, this is a necessary but not a sufficient condition since there are certain classes of closed loops which will not cause an instability, cf. problem 4. A very interesting example for an A1-network which really shows instability is the case in which two A1-elements are arranged in parallel such that they originate with their input ports as branches from a 0-junction and are connected at their output ports by a 1-junction:

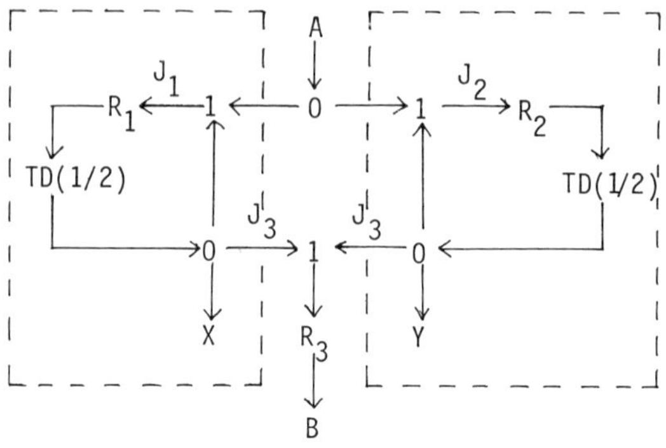

(6.16)

This network is particularly interesting for vanishingly small concentrations of the external species B such that with the usual product ansatz for the fluxes along R_1, R_2 and R_3 we obtain

$$\left.\begin{aligned} \frac{dX}{dt} &= J_1 - J_3 = (k_1A - k_1'X - k_3Y)X \equiv \varphi(X,Y)\cdot X \\ \frac{dY}{dt} &= J_2 - J_3 = (k_2A - k_2'Y - k_3X)Y \equiv \psi(X,Y)\cdot Y \end{aligned}\right\} . \qquad (6.17)$$

For the steady state of the network we have the following possibilities: (A) $\bar{X} = \bar{Y} = 0$; (B1) $\varphi(\bar{X},\bar{Y}) = 0$, $\bar{Y} = 0$; (B2) $\psi(\bar{X},\bar{Y}) = 0$, $\bar{X} = 0$; (C) $\varphi(\bar{X},\bar{Y}) = 0$, $\psi(\bar{X},\bar{Y}) = 0$. In order to decide which one of the various steady states is stable we consider the configurations of the signs of dX/dt and dY/dt in the X-Y-plane. At $\varphi(X,Y) = 0$ which defines a straight line in the X-Y-plane, dX/dt changes its sign and likewise dY/dt at $\psi(X,Y) = 0$. The parameters which will determine the stability behaviour are $\lambda_1 = k_1k_2'/k_2$, $\lambda_2 = k_2k_1'/k_1$ and k_3. Without restriction let us assume $\lambda_1 \geq \lambda_2$. The steady states and the sign configurations are presented in Fig. 12.

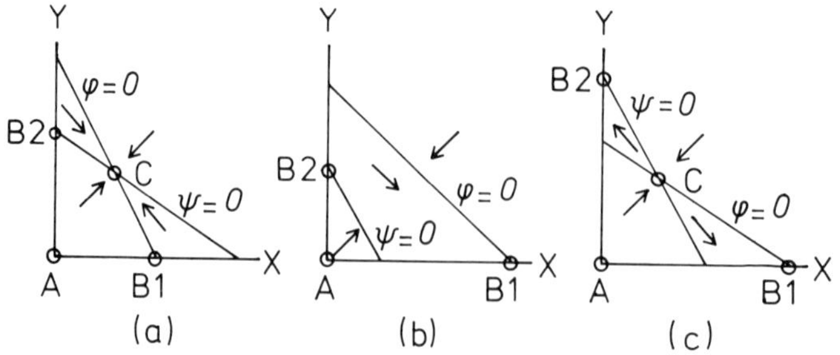

Fig. 12. Sign configurations of dX/dt and dY/dt of the network (6.16), (a) for $k_3 < \lambda_2$, (b) for $\lambda_2 < k_3 < \lambda_1$ and (c) for $\lambda_1 < k_3$

The arrows in Fig. 12 result from combining the signs of dX/dt and dY/dt and thus indicate the momentary direction of evolution of the state of the system. From Fig. 12 we conclude that for $k_3 < \lambda_2$ the steady state (C) is globally and asymptotically stable whereas (A), (B1) and (B2) are instable. For $\lambda_2 < k_3 < \lambda_1$, the state (C) is absent since the solution of the system of linear equations $\varphi(\overline{X},\overline{Y}) = 0$, $\psi(\overline{X},\overline{Y}) = 0$ would lead to negative values for $\overline{X},\overline{Y}$ which is physically unreal. In this case, (B1) is globally and asymptotically stable whereas (A) and (B2) are instable. The situation changes essentially, however, if $\lambda_1 < k_3$, since now both (B1) and (B2) are locally and asymptotically stable, the local character of stability necessarily following from the simultaneous appearance of two stable steady states. As before, (A) is instable whereas (C) has a saddle-point character, i.e., stability and instability for particular directions, which clearly implies instability on the whole.

Our model can be given a direct ecological meaning by interpreting X and Y as self-reproducing species with an interspecific aggression $X + Y \longrightarrow B$ expressed by the reaction R_3 in (6.16), cf. MacARTHUR (1970), MAY (1973). With this interpretation, the steady state (C) represents the coexistence of the two species whereas (A), (B1), (B2) are ecologically degenerate in that at least one species will die out. The model now predicts that coexistence is possible only for subcritical interspecific aggression $k_3 < k_{3c} = \lambda_2$ whereas for supercritical values $k_3 > k_{3c}$ only one of the species will survive. Surprisingly, the surviving species is uniquely determined in the interval $\lambda_2 < k_3 < \lambda_1$ but undetermined if the interspecific aggression increases up to values of $\lambda_1 < k_3$. In the latter case, the system is physically degenerate in that two steady states appear very similarly as in the cooperative models of the Ising type in Section 2.5 and of the dissipative autocatalytic type in Section 6.2. Particularly the symmetric case $\lambda_1 = \lambda_2$ resembles very much the situation of a phase transition as k_3 passes the critical value $k_{3c} = \lambda_1 = \lambda_2$.

6.4 Limit Cycle Networks

The instabilities which we have generated by feedback loops in the networks of the preceding sections of this chapter are all of the type of multiple steady states. In this section we shall discuss the problem of how to construct networks which have an instability of a limit cycle type. From Section 2.7 we already know a model which shows this kind of instability, namely, a feedback model for the control of metabolic reaction chains. Following the network rules which we have developed so far, the network representation of this model is given as

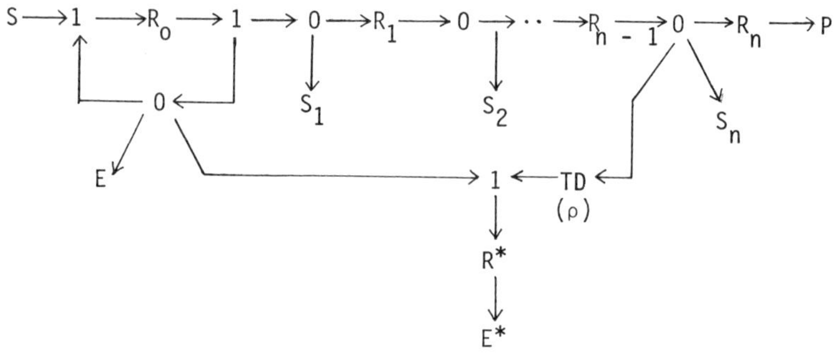

(6.18)

We have mentioned already in Section 2.7, and this fact is also immediately confirmed by inspection of the structure of the network in (6.18), that the loop in (6.18) which causes the instability effectively acts as a negative feedback. This might lead to the suspicion that quite generally positive feedback causes multiple steady states as was the case in Sections 6.1, 6.2 and 6.3 whereas negative feedback causes limit cycles as in (6.18). We shall see, however, that things are not that simple.

Let us consider the appearance of limit cycles from a little more systematic point of view. It is clear that a network which shows limit cycles necessarily has at least two variables since a real first-order differential equation in one single variable never can produce periodic solutions. Let X_1, X_2 be the concentration variables of the network and let their equations of motion be given as

$$\frac{dX_i}{dt} = \Phi_i(X_1,X_2) \qquad i = 1,2 \quad . \tag{6.19}$$

Assume that we have found a steady state $\overline{X}_1$, $\overline{X}_2$ such that $\Phi_i(\overline{X}_1, \overline{X}_2) = 0$ for $i = 1,2$. Let us perform a differential analysis in the vicinity of the steady state such that we obtain from (6.19)

$$\frac{d}{dt}\,\delta X_i = \Phi_{i1}\delta X_1 + \Phi_{i2}\delta X_2 \qquad i = 1,2 \tag{6.20}$$

where Φ_{ij} is the derivative of Φ_i with respect to X_j taken at the steady state. Defining now new variables u_1, u_2 by

$$u_1 = \delta X_1 \quad , \qquad u_2 = \Phi_{11}\delta X_2 + \Phi_{12}\delta X_2 \tag{6.21}$$

the system (6.20) is readily transformed into

$$\dot{u}_1 = u_2 \quad , \qquad \dot{u}_2 = - \Delta \cdot u_1 + 2\lambda u_2 \tag{6.22}$$

where the dot indicate the time derivative and

$$\Delta = \Phi_{11}\Phi_{22} - \Phi_{12}\Phi_{21} \quad , \qquad \lambda = - \frac{1}{2} (\Phi_{11} + \Phi_{22}) \quad . \tag{6.23}$$

Taking the second-order time derivative of u_1 we immediately obtain

$$\ddot{u}_1 + 2\lambda\dot{u}_1 + \Delta \cdot u_1 = 0 \tag{6.24}$$

which has the form of a damped harmonic oscillator if $\Delta > 0$ and $\lambda > 0$. In this case, $\omega_0 = \pm\sqrt{\Delta}$ is the frequency of the oscillator and λ is its damping rate such that the amplitude is decreasing with time as exp $(-\lambda t)$.

As in Section 2.7 we now formulate as a condition for the onset of the limit cycle that the oscillator in (6.24) is not damped but amplified, i.e., $\lambda < 0$, but still $\Delta > 0$. We expect that λ as defined in (6.23) is some function of the external parameters of the network such that at least near the thermodynamic equilibrium we have $\lambda > 0$ and only for sufficiently strong nonequilibrium conditions for the external parameters λ becomes negative. This behaviour is typically reflected by the conditions (2.75) and (2.76) for our model in Section 2.7. At the critical point, the so-called bifurcation point, which separates the damped region from the amplifying one, we have $\lambda^c = 0$ and thus from the definition of λ in (6.23) opposite signs of Φ_{11}^c and Φ_{22}^c such that $\Phi_{11}^c \cdot \Phi_{22}^c < 0$. Since on the other hand we still have to fulfill $\Delta^c > 0$ we conclude from the definition of Δ in (6.23) that necessarily $\Phi_{12}^c \cdot \Phi_{21}^c < 0$. We can now think of two possibilities to satisfy the combined condition

$$\Phi_{11}^c\Phi_{22}^c < 0 \quad , \qquad \Phi_{12}^c\Phi_{21}^c < 0 \tag{6.25}$$

namely (A) $\Phi_{22}^c > 0$, $\Phi_{21}^c > 0$ and (B) $\Phi_{22}^c < 0$, $\Phi_{21}^c > 0$. All other cases follow from these two by permutation of the indices. The case (A) can be realized by rather simple networks. It evidently means an autocatalytic production of X_2 at the expense of X_1 since both $\delta X_1 > 0$ and $\delta X_2 > 0$ cause in increase of $d\delta X_1/dt$. A system which follows this line of construction is the so-called "Brusselator", cf. GLANSDORFF, PRIGOGINE (1971), TYSON (1973):

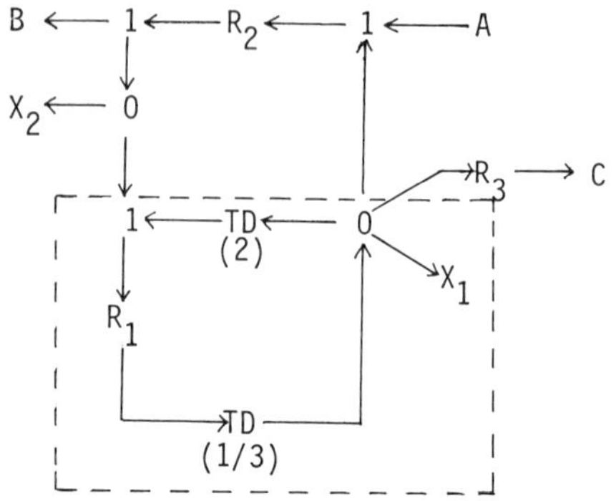

. (6.26)

First of all we observe that the network (6.26) contains an A2-element as integrated circuit for the autocatalytic production of X_1. In addition to this autocatalytic loop there is a second loop formed by the reaction R_2. From the topological point of view, there is a strikingly close resemblance between the Brusselator in (6.26) and our autocatalytic excitation model in (6.7). Despite this topological resemblance, the two networks behave fundamentally differently, namely multiple steady states in (6.7) and a limit cycle in (6.26). This shows that a purely network topological classification of types of models will find its limits or is at least a pretentious program.

Another realization of the above case (A) i.e., $\Phi_{22}^{c} > 0$, $\Phi_{21}^{c} > 0$ has been proposed by BALSLEV and DEGN (1975):

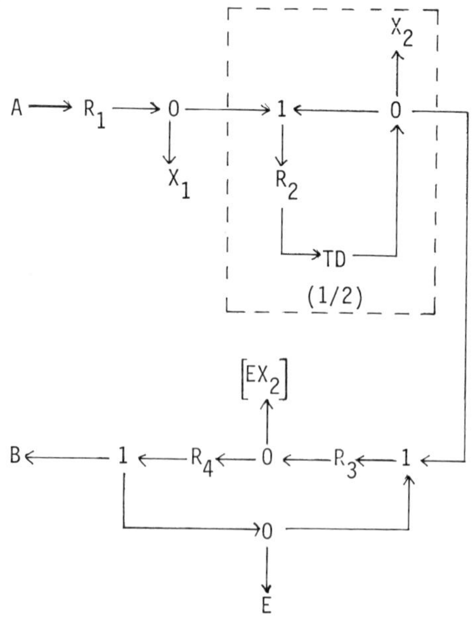

. (6.27)

This network manages its purpose, namely a limit cycle, with an A1-element instead of an A2-element in (6.26) and it also avoids a second loop as needed in (6.26). On the other hand, it introduces two new variables, namely an enzyme E and the enzyme-substrate complex $\left[EX_2\right]$ in the manner of a Michaelis-Menten kinetics as in Section 2.2. Following the treatment of Balslev and Degn, the reader may readily evaluate the limit cycle condition of (6.25) by assuming that the enzyme-catalyzed reaction steps R_3 and R_4 are very rapid compared to those of R_1 and R_2 and that the whole system is so far from equilibrium that only the forward rates of R_1 and R_2 leading from A to B have to be taken into account (cf. problem 6).

Concerning further biological applications of limit cycles, there is by now an immense literature on this subject. The interested reader is referred to two review articles by NICOLIS and PORTNOW (1973) and NOYES and FIELD (1974) where further literature is cited.

Problems

1) Show that choices of ν other than $\nu = 1,2$ in (6.3) and even independent choices of the moduli of the two transducers in (6.3) do not lead to elements which are qualitatively different from A1 and A2.
2) Decompose the autocatalytic reaction system in (6.3) into partial steps by introducing intermediate reaction products, e.g., for $\nu = 2$

$$A + X \rightleftharpoons Y \quad , \qquad Y \rightleftharpoons 2X \quad , \qquad X \rightleftharpoons B$$

and for $\nu = 3$

$$A + X \rightleftharpoons Y \quad , \qquad Y + X \rightleftharpoons Z + X \quad , \qquad Z \rightleftharpoons 2X \quad , \qquad X \rightleftharpoons B \quad .$$

Find other decompositions and construct the corresponding networks. Show that the essential global behaviour remains qualitatively unchanged under such decompositions.
3) Plot dX/dt of the Ising model in (6.14) as a function of X similarly as in Fig. 10 and prove that in the case of three steady states $\bar{X}_1 < \bar{X}_2 < \bar{X}_3$ $\bar{X}_2$ is instable whereas $\bar{X}_1$ and $\bar{X}_3$ are locally and asymptotically stable. What are the local stability regions of $\bar{X}_1$ and $\bar{X}_3$?
4) Construct and discuss the Volterra-Lotka network for the case that a self-reproducing prey X_1 has two competing predators X_2, X_3 which in turn are pursued by a common predator X_4.

5) Consider the network (6.16) for nonvanishing values of the external concentration of B and discuss the limit $B \longrightarrow 0$. How are the stability properties changed for $B \neq 0$?

6) BALSLEV and DEGN (1975) have given further examples of limit cycle systems in their paper. Design the corresponding networks, look for integrated circuits like A1 and A2 and evaluate the differential limit cycle condition $\lambda < 0$ and $\Delta > 0$ of (6.23).

7) Modify the model for the control of metabolic reaction chains as presented by the network in (6.18) in such a way that the enzyme E is inactivated by a specific blocker B, $E + B \rightleftharpoons E^*$, which in turn is inactivated by ρ molecules of the intermediate product S_n, $B + \rho S_n \rightleftharpoons B^*$. Construct the network for the modified model and argue that the feedback character has been changed by this modification from negative to positive. Show that the modified model has a multiple steady state similar to the networks with a direct positive feedback. Is the modified model biologically meaningful? (Actually, this modification is the essential core of LAVENDA's (1972) model for the control of biosynthesis of proteins.)

7. Stability

7.1 Capacities as Thermodynamic Equilibrium Systems

Particularly in the preceding Chapter 6 we have seen the significance of stability properties for biological network models. This experience leads us to devote the present chapter to a development of quite general techniques for a stability analysis of networks. Such an analysis can be performed under two different aspects: a thermodynamic aspect, which relates stability to thermodynamic properties of the network like entropy production and the second law; and a nonthermodynamic aspect, which derives the stability properties from the mathematical structure of the differential equations represented by the network on the basis of a topological analysis. Sections 7.1 to 7.4 of this chapter will be devoted to the thermodynamic aspect and in Sections 7.5 to 7.7 we briefly describe a few simple techniques to obtain information on stability from a topological analysis.

In the present section, we start our thermodynamic stability considerations by recalling and generalizing the definition of capacitances as introduced in Section 4.2. Quite generally, capacitances are the equilibrium parts of thermodynamic networks or, in other words, they are reversible storage elements for extensive quantities like energy U, volume V, and mole numbers N_1, N_2, ... The property of the reversibility of the storage process is expressed by Gibbs' relation which in its version (3.65) for the volume densities $s = S/V$, $u = U/V$, $c_i = N_i/V$ can be written as

$$\delta s = \frac{1}{T}\,\delta u - \sum_i \frac{\mu_i}{T}\,\delta c_i \tag{7.1}$$

or for processes continuous in time as

$$\dot{s} = \frac{1}{T}\,J_u - \sum_i \frac{\mu_i}{T}\,J_i \tag{7.2}$$

with

$$J_u = \frac{du}{dt}\ , \qquad J_i = \frac{dc_i}{dt}\ . \tag{7.3}$$

The reversibility of storage is reflected by the fact that δs in (7.1) or $\dot{s}$ in (7.2) can be positive or negative depending on the combinations of the signs of δu, δc_i or J_u, J_i, respectively.

If the system contains a number of n different kinds of molecules, i.e., $1 \leq i \leq n$, Gibbs' relation in its version of (7.2) is represented now by a (n + 1)-port capacitance element where each port is characterized by a pair of variables (-1/T, J_u) or (μ_i/T, J_i), $1 \leq i \leq n$:

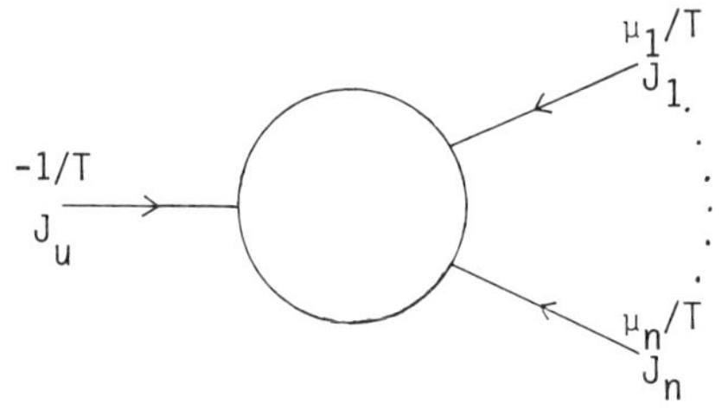

. (7.4)

The port variables in (7.4) are pairs of an intensive variable -1/T or μ_i/T and a flux variable J_u or J_i. Let us call the intensive variables -1/T and μ_i/T the potentials of the ports. Potentials and fluxes of the ports are then related to each other by

$$-\frac{1}{T} = -\frac{\delta \dot{s}}{\delta J_u} \quad , \quad \frac{\mu_i}{T} = -\frac{\delta \dot{s}}{\delta J_i} \tag{7.5}$$

or vice versa.

Now the concept of the capacitance matrix as introduced already in (4.6), (4.7) is brought into play be writing the densities u and c_i as functions of the potentials,

$$u = u\left(-\frac{1}{T}, \frac{\mu_1}{T}, \dots \frac{\mu_n}{T}\right) \quad , \quad c_i = c_i\left(-\frac{1}{T}, \frac{\mu_1}{T}, \dots \frac{\mu_n}{T}\right) \tag{7.6}$$

such that

$$\left.\begin{aligned} J_u &= \frac{du}{dt} = C_{uu}\frac{d}{dt}\left(-\frac{1}{T}\right) + \sum_j C_{uj}\frac{d}{dt}\left(\frac{\mu_j}{T}\right) \\ J_i &= \frac{dc_i}{dt} = C_{iu}\frac{d}{dt}\left(-\frac{1}{T}\right) + \sum_{i,j} C_{ij}\frac{d}{dt}\left(\frac{\mu_j}{T}\right) \end{aligned}\right\} \tag{7.7}$$

where

$$\left.\begin{array}{ll} C_{uu} = \dfrac{\partial u}{\partial(-1/T)} & C_{uj} = \dfrac{\partial u}{\partial(\mu_j/T)} \\[2ex] C_{iu} = \dfrac{\partial c_i}{\partial(-1/T)} & C_{ij} = \dfrac{\partial c_i}{\partial(\mu_j/T)} \end{array}\right\} \qquad (7.8)$$

Eq. (7.7) expresses the fluxes into the (n + 1)-port capacitance as linear combinations of the time derivates of the potentials and thus represents the constitutive equation of the capacitance. If the capacitance matrix in (7.8) is diagonal, i.e., $C_{uj} = 0$, $C_{iu} = 0$ and $C_{ij} = 0$ for $i \neq j$, (7.4) decomposes into independent 1-port capacitances.

If the real thermodynamic system to be represented by a network is spatially homogeneous we have to introduce but one single capacitance for the storage processes of the system. Chemical reactions occurring in the system are described then by 2-port reaction elements which join different ports of the capacitance. Under special conditions, e.g., in ideal solutions (see below), it may happen that the (n + 1)-port capacitance decomposes into separate 1-port capacitances. If, however, the real system is spatially inhomogeneous, we divide the total system into spatial subsystems $\alpha = 1,2, \ldots$ such that each subsystem can be treated at least approximately as a thermodynamic equilibrium system. Now, for each of the subsystems $\alpha = 1,2, \ldots$ separate (n + 1)-port capacitances are introced and joined together by 2-port diffusion elements as introduced in Section 4.6 and possibly also by quite similar 2-port elements for heat conduction if the subsystems have different temperatures. As before, the 2-port reaction elements are placed between the ports of the capacitance of a single subsystem α. Evidently, capacitances are the network counterparts of the principle of local equilibrium in continuous systems.

Since entropy is an extensive quantity its total variation δS or its total time derivative $\dot{S}$ is obtained by summing the variations or time derivatives of the entropies of the subsystems $\alpha = 1,2, \ldots$. Recalling that (7.2) expresses the time derivative of the volume density of the entropy we thus have

$$\dot{S} = \sum_{\alpha} V_{\alpha} \left(\frac{1}{T_{\alpha}} J_{u\alpha} - \sum_{i} \frac{\mu_{i\alpha}}{T_{\alpha}} J_{i\alpha} \right) \qquad (7.9)$$

provided that the volumes V_{α} of the subsystems remain fixed during the processes within the network. Clearly, this latter assumption excludes convection processes. T_{α}, $\mu_{i\alpha}$, $J_{u\alpha}$ and $J_{i\alpha}$ are the temperature, the chemical potentials and the fluxes of energy and molecules of kind i into the capacitance of subsystem α.

A very important generalization of the concept of capacitances is concerned with a network coupled to bath systems for energy or matter. A bath system can be considered as a 1-port capacitance with a constant potential variable even for nonvanishing fluxes along the port. For example, a heat bath is a system with constant temperature T_0 although it exchanges finite amounts $V_0 J_u$ of energy with its surrounding, V_0 being the volume of the heat bath. Applying (7.7) to this particular situation we may write

$$V_0 J_u = V_0 C_0 \frac{d}{dt}\left(-\frac{1}{T_0}\right) \tag{7.10}$$

where C_0 is the capacity of the density of energy and $V_0 C_0$ is the total capacity of energy of the bath. From (7.10) we conclude that constant temperature T_0 at nonvanishing $V_0 J_u$ is formally reflected by $V_0 C_0 \longrightarrow \infty$ which just expresses that bath systems are very large systems.

Let us assume now that the α-sum in (7.9) includes the capacitances of all bath systems for energy and matter to which the network is coupled. With this convention we can now treat the whole system as isolated such that $\dot{S} \geq 0$ in (7.9), i.e., $\dot{S} > 0$ for spontaneous processes of the network and $\dot{S} = 0$ defines the equilibrium of the network.

Apart from the assumption of fixed volumes of the subsystems, $dV_\alpha/dt = 0$, our treatment of thermodynamic capacitances has been quite general so far. For the following considerations in this chapter let us restrict ourselves to isothermal networks by coupling each of the capacitances of the network to a heat bath with given and fixed temperature T. This immediately reduces (7.9) to

$$\dot{S} = \frac{1}{T}\sum_\alpha V_\alpha \left(J_{u\alpha} - \sum_i \mu_{i\alpha} J_{i\alpha}\right) \quad . \tag{7.11}$$

Since the complete system is isolated, we have

$$\sum_\alpha V_\alpha J_{u\alpha} = 0 \tag{7.12}$$

and thus

$$\dot{S} = -\frac{1}{T}\sum_{\alpha,i} V_\alpha \mu_{i\alpha} J_{i\alpha} \quad . \tag{7.13}$$

Eq. (7.12) means that we can omit the energy ports and bonds in our networks. In (7.13) the α-summation runs through all subsystems and the i-summation runs through all kinds of molecules which are present in the system. Let us simplify the summation notation by introducing one single summation index i which runs through all ports of all capacitances such that (7.13) becomes

$$\dot{S} = -\frac{1}{T}\sum_i V_i \mu_i J_i = -\frac{1}{T}\sum_i \mu_i I_i \tag{7.14}$$

where

$$I_i = V_i J_i = V_i \frac{dc_i}{dt} = \frac{dN_i}{dt} \tag{7.15}$$

and V_i denotes the volume of the capacitance to which the port with number i is connected. Eq. (7.14) makes it suggestive to redefine the potentials of the ports and the capacitance matrix by writing (at constant temperature T)

$$J_i = \frac{d}{dt} c_i\ (\mu_1, \dots \mu_n) = \sum_j C_{ij} \frac{d\mu_j}{dt} \tag{7.16}$$

as the constitutive relation for capacitances such that μ_i instead of μ_i/T is the potential of the port i and

$$C_{ij} = \left(\frac{\partial c_i}{\partial \mu_j}\right)_{T,\mu_k} \tag{7.17}$$

is the capacitance matrix. Whether I_i or J_i is defined as the flux of the port i is just a question of convenience since the volumes V_i had been assumed to be kept fixed. From

$$\delta f = -s\delta T + \sum_i \mu_i \delta c_i \tag{7.18}$$

where f = F/V is the volume density of the free energy (cf. problem 2 of Chapter 3) we derive

$$\frac{\partial \mu_i}{\partial c_j} = \frac{\partial^2 f}{\partial c_j \partial c_i} = \frac{\partial^2 f}{\partial c_i \partial c_j} = \frac{\partial \mu_j}{\partial c_i} \quad . \tag{7.19}$$

Eq. (7.19) expresses that the matrix $\partial\mu_i/\partial c_j$ is symmetric. Since C_{ij} is the inverse of $\partial\mu_i/\partial c_j$ we have thus shown that the capacitance matrix C_{ij} is symmetric: $C_{ij} = C_{ji}$. By the very definition of C_{ij} we also know that $C_{ij} = 0$ if ports i and j belong to different capacitances. If the system is an ideal solution such that the relationship between the concentrations and chemical potentials is given by (3.77) the capacitance matrix is diagonal and all capacitances decompose into 1-port capacitances. With these redefinitions we have returned to the preliminary definition of a capacitance in the potential-flux language as given in Section 4.2.

Finally let us reformulate (7.14) and (7.16) in terms of differential variations:

$$\delta S = -\frac{1}{T}\sum_i V_i\mu_i\delta c_i \quad , \qquad \delta c_i = \sum_j C_{ij}\delta\mu_j \quad . \tag{7.20}$$

From (7.20) we immediately derive

$$\delta S = -\frac{1}{T}\sum_{i,j} C_{ij}V_i\mu_i\delta\mu_j \quad . \tag{7.21}$$

Since $C_{ij} = 0$ if ports i and j are connected to different capacitances and $V_i = V_j$ if ports i and j are connected to the same capacitance the matrix $K_{ij} = V_iC_{ij}$ is symmetric and we may write

$$\delta S = -\frac{1}{T}\sum_{i,j} K_{ij}\mu_i\delta\mu_j \quad . \tag{7.22}$$

As mentioned already earlier in this section we have $\dot{S} > 0$ or $\delta S > 0$ for spontaneous processes and $\dot{S} = 0$ or $\delta S = 0$ as definitions for the equilibrium.

7.2 Stability of the Equilibrium State

As argued in the preceding section the thermodynamic equilibrium of the network is characterized by a vanishing first-order differential variation of its total entropy: $\delta S = 0$. Due to (7.14) or (7.21) δS is given as

$$\delta S = -\frac{1}{T}\sum_i V_i\mu_i\delta c_i = -\frac{1}{T}\sum_{i,j} C_{ij}V_i\mu_i\delta\mu_j \quad . \tag{7.23}$$

Evaluating $\delta S = 0$ for all independent variations of δc_i or $\delta\mu_i$, i.e., for all variations which satisfy the internal constraints of the network as expressed by KCL and KVL at the 0- and 1-junctions, leads to the determination of the equilibrium values $\overline{\mu}_i^e$ of the chemical potentials. This is exactly the way we have determined equilibrium states in Sections 3.4, 3.5, and 3.6.

Entropy $\delta S = 0$ is a necessary, but not a sufficient condition. For a stable thermodynamic equilibrium state we really need a maximum of S and not just an extremum or a saddle-point which both could still be included in $\delta S = 0$. In the case of a saddle-point, for example, the system would spontaneously leave the equilibrium and follow a direction of increasing entropy S due to the second law of thermodynamics. A sufficient condition for a maximum of S at the equilibrium state characterized by the potentials $\overline{\mu}_i^e$ (at given and fixed temperature T) is thus a negative second-order variation $(\delta^2 S)^e < 0$ of the total network entropy S. In order to draw further conclusions from this criterion we first have to calculate the second-order variation $\delta^2 S$ and then to evaluate it at the equilibrium as defined by the equilibrium values $\overline{\mu}_i^e$. The first step is easily performed by applying the variational operation δ to the first expression for δS in (7.23). Considering the concentrations c_i as independent variables, i.e., $\delta^2 c_i = 0$, we obtain by applying the product rule of differentiation

$$\delta^2 S = -\frac{1}{T}\sum_i V_i \delta\mu_i \delta c_i \tag{7.24}$$

for fixed values of the temperature T and the volume V_i. Returning now from the c_i to the potentials μ_i as independent variables and expressing δc_i in terms of $\delta\mu_i$ by making use of (7.20) we obtain

$$\delta^2 S = -\frac{1}{T}\sum_{i,j} C_{ij} V_i \delta\mu_i \delta\mu_j \quad . \tag{7.25}$$

The sufficient condition for the stability of the equilibrium now reads $(\delta^2 S)^e < 0$ or

$$-T(\delta^2 S)^e = \sum_{i,j} C_{ij}^e V_i \delta\mu_i \delta\mu_j = \sum_{i,j} K_{ij}^e \delta\mu_i \delta\mu_j > 0 \tag{7.26}$$

for all independent differential variations $\delta\mu_i = \mu_i - \overline{\mu}_i^e$ around the equilibrium state $\overline{\mu}_i^e$. The superscript in C_{ij}^e and K_{ij}^e indicates that C_{ij} of (7.17) has to be evaluated at the equilibrium state $\overline{\mu}_i^e$. The expression $-T(\delta^2 S)^e$ as given by (7.26) is a so-called quadratic form. If this form is positive definite, or in other words, if the capacitance matrix C_{ij}^e is a positive definite matrix, the thermodynamic equilibrium is stable. If the capacitance matrix C_{ij}^e is diagonal, i.e., if each of the concentrations c_i depends only on its own chemical potential μ_i as for example in dilute solutions, the multi-port capacitances decompose into 1-port capacitances and (7.26) is reduced to

$$-T(\delta^2 S)^e = \sum_i V_i \left(\frac{dc_i}{d\mu_i}\right)^e \left(\delta\mu_i\right)^2 > 0 \quad . \tag{7.27}$$

Eq. (7.27) requires the concentrations c_i to be monotonically increasing functions of their potentials μ_i: the higher the value of the concentration, the higher its chemical potential. Although this property seems to be self-evident there are situations in which $-T(\delta^2S)^e > 0$ is not satisfied, e.g., for phase transitions of the kind as decribed by the Ising model in Section 2.5 where in the coexistence region there may be an exchange $\delta c_i \neq 0$ between the two phases at constant potential, $\delta\mu_i = 0$, and hence $\delta^2S = 0$ from (7.24). This means that the second-order stability criterion would predict an indifferent equilibrium state. It is clear that in such a case one would have to investigate the higher-order variations δ^3S, δ^4S, ...

For the purpose of later applications to nonequilibrium states we now generalize the formulation the second-order stability criterion in (7.26). Interpreting $-T(\delta^2S)^e$ in (7.26) as a function of the variations $\delta\mu_i$ around the fixed equilibrium state we calculate its time derivative by making use of the symmetry of the matrix K_{ij}:

$$\frac{d}{dt}\left(-T(\delta^2S)^e\right) = 2\sum_{i,j} K_{ij}^e \delta\mu_i \frac{d}{dt}\delta\mu_j = 2\sum_i V_i \delta\mu_i \frac{d}{dt}\delta c_i = 2\sum_C V_i \delta\mu_i \delta J_i \quad . \tag{7.28}$$

In the last step of (7.28), $d\delta c_i/dt$ has been replaced by the variations of the capacitive flux J_i for the associated reference direction, cf. (4.1), and the sum over i has been marked by the symbol C in order to recall that i runs through all ports of all capacitances in the network. Defining

$$L = -\frac{1}{2}(\delta^2S)^e \tag{7.29}$$

we can now formulate the stability of the equilibrium as a so-called Liapunov-criterion with L as Liapunov-function: if L is a nonnegative function, $L \geq 0$, such that $L = 0$ is valid only in the equilibrium state, and if $dL/dt \leq 0$ such that again $dL/dt = 0$ is valid only in the equilibrium, then it is evident that the system necessarily tends into the equilibrium state as $t \longrightarrow \infty$ and that it never can leave this state again after having arrived at it. Liapunov-criteria can be formulated under much less restrictive conditions, cf. LASALLE and LEFSHETZ (1961). For our purposes, the above formulation is sufficiently general. The condition $L \geq 0$ is identical with that of a positive definite capacitance matrix whereas $dL/dt \leq 0$ has been considered as a consequence of the second law of thermodynamics. It is worth while reinvestigating this latter point with some more care particularly in view of later applications to the stability of nonequilibrium states. This will be done by making use of a typical network theorem in the following section.

7.3 Tellegen's Theorem and Liapunov Stability of the Equilibrium

The main point of this section is the proof of a very general network theorem which is frequently referred to as Tellegen's theorem (TELLEGEN 1952). It says that

$$\sum_E V_i\mu_i J_i = \sum_E \mu_i I_i = 0 \tag{7.30}$$

where the symbol E indicates that the summation variable i runs through the ports of all elements of the network, i.e., including capacitances and the dissipative elements like reaction 2-ports, etc. Moreover, let us choose the total fluxes $I_i = V_i J_i$ as the flux variables of the ports for the considerations in this section and let us assume that all ports are chosen with associated reference directions: $I_i > 0$ if the flux is directed into the network element with port i.

To prove (7.30), we first assume that all elements are 1-ports and afterwards generalize the result. Let 0_s, s = 1,2, ... N_0, and 1_t, t = 1,2, ... N_1 denote the 0- and 1-junctions of the network, respectively, and let $\varepsilon_s = 1,2, \ldots \nu_s$ and $\varepsilon_t = 1,2, \ldots \nu_t$ characterize the elements which are connected with 0_s or 1_t, respectively. With this notation we can decompose (7.30) to

$$\sum_E \mu_i I_i = \sum_{s=1}^{N_0} \sum_{\varepsilon_s = 1}^{\nu_s} \mu_{\varepsilon_s}^{(s)} I_{\varepsilon_s}^{(s)} + \sum_{t=1}^{N_1} \sum_{\varepsilon_t = 1}^{\nu_t} \mu_{\varepsilon_t}^{(t)} I_{\varepsilon_t}^{(t)} \tag{7.31}$$

where $\mu_{\varepsilon_s}^{(s)}$ and $I_{\varepsilon_s}^{(s)}$ denote the potential and the flux along the bond which connects an element to 0_s and similarly $\mu_{\varepsilon_t}^{(t)}$ and $I_{\varepsilon_t}^{(t)}$. Now we express Kirchhoff's current (KCL) and voltage (KVL) law as

$$\text{KCL:} \quad \sum_{\varepsilon_s = 1}^{\nu_s} I_{\varepsilon_s}^{(s)} + \sum_{t=1}^{N_1} \sigma_{st} I_{st} = 0 \tag{7.32}$$

$$\text{KVL:} \quad \sum_{\varepsilon_t = 1}^{\nu_t} \mu_{\varepsilon_t}^{(t)} + \sum_{s=1}^{N_0} \sigma_{ts}\mu_{ts} = 0 \tag{7.33}$$

where σ_{st} = +1 or -1 if 0_s is connected to 1_t with reference direction from 0_s to 1_t or vice versa, σ_{st} = 0 otherwise, and similarly σ_{ts}. I_{st} and μ_{ts} are the flux and the potential along the bond between 0_s and 1_t if existing. There are no contributions from bonds between different 0-junctions to (7.20) or from bonds between different 1-junctions to (7.21) since directly connected 0- or 1-junctions can always be contracted into one single junction without loss of generality, cf. problem 1 of Chapter 4. It should also be noted in this context that KCL in its most general form has to be expressed actually in terms of the total fluxes $I_i = dN_i/dt$ instead of the fluxes $J_i = dc_i/dt$ since KCL expresses a conservation law for an extensive quantity and not necessarily for its densities. Only if we choose all volumes V_i of the subsystems as being equal as we did in the network models in the preceding chapters KCL may equivalently be written in terms of the J_i.

Remembering now $\mu_1^{(s)} = \mu_2^{(s)} = \ldots \equiv \mu_s$ at 0_s and $I_1^{(t)} = I_2^{(t)} = \ldots \equiv I_t$ at 1_t, cf. (4.12) and (4.30), we obtain upon inserting (7.32) and (7.33) into (7.31)

$$\sum_E \mu_i I_i = -\sum_{s,t} (\mu_s \sigma_{st} I_{st} + I_t \sigma_{ts} \mu_{ts}) = -\sum_{s,t} (\mu_s I_{st} - \mu_{ts} I_t) \sigma_{st} = 0 \tag{7.34}$$

since $\sigma_{st} = -\sigma_{ts}$ and either $\sigma_{st} = 0$ or $I_{st} = I_t$ and $\mu_{ts} = \mu_s$ if $\sigma_{st} \neq 0$.

Our proof is immediately extended to include networks with arbitrary multi-port elements. At no point of the proof have we made use of the assumption that two different ports really belong to two different elements. We just have to bear in mind that a direct connection between two ports is impossible since we have assumed associated reference directions for all ports. This means that every port of a 1- or multi-port element is necessarily connected either to a 0- or to a 1-junction.

Another very important point of the above proof is the fact that the only prerequisites which have really been made use of are KCL and KVL but not the physical state of the network neither any other thermodynamic property. In other words, Tellegen's theorem is a purely topological theorem. This means that (7.30) remains valid even if μ_i and I_i are interpreted as the values of the potentials and fluxes at different times which in turn implies

$$\sum_E \mu_i \frac{dI_i}{dt} = 0 \quad , \qquad \sum_E \frac{d\mu_i}{dt} I_i = 0 \tag{7.35}$$

or even

$$\sum_E \frac{d\mu_i}{dt} \cdot \frac{dI_i}{dt} = 0 \quad . \tag{7.36}$$

Eq. (7.36) can equivalently be written as

$$\sum_E \delta\mu_i \delta I_i = \sum_E V_i \delta\mu_i \delta J_i = 0 \tag{7.37}$$

for the variations around any state of the system, in particular around the equilibrium. Decomposing the sum in (7.37) into a C-part over all capacitances and a R-part over all dissipative elements we obtain

$$\sum_C V_i \delta\mu_i \delta J_i + \sum_R V_i \delta\mu_i \delta J_i = 0 \qquad (7.38)$$

which upon insertion into (7.28) together with the definition (7.29) gives

$$\frac{dL}{dt} = \frac{d}{dt}\left[-\frac{1}{2}(\delta^2 S)^e\right] = -\frac{1}{T}\sum_R V_i \delta\mu_i \delta J_i \quad . \qquad (7.39)$$

Let us apply now (7.39) to networks where the dissipative elements are reaction or diffusion 2-ports as was the case in the models of Chapters 5 and 6. For a reaction 2-port of the type

$$\overset{A^f}{\underset{J}{\longrightarrow}} R \overset{A^r}{\underset{J}{\longrightarrow}} \qquad (7.40)$$

cf. (4.27) and (4.28), the corresponding contribution to the R-sum in (7.39) is given by

$$\frac{1}{T}\left[\delta A^f \cdot \delta J + \delta A^r \cdot (-\delta J)\right] = \frac{1}{T}\,\delta A \cdot \delta J \qquad (7.41)$$

since the port on the reverse side of (7.40) has nonassociated reference orientation. $A = A^f - A^r$ is the over-all affinity of the reaction, cf. (4.24). Quite a similar transformation is performed in the case of a diffusion 2-port such that by comparison of (7.39) or (7.41) with the definition of generalized thermodynamic forces and fluxes in (3.79) we eventually may write

$$\frac{dL}{dt} = -\sum_\alpha \delta F_\alpha \cdot \delta I_\alpha \qquad (7.42)$$

where α runs through all dissipative 2-port elements of the network.

Finally we recall that due to (7.27) the variations in (7.42) are to be interpreted as differential variations around the thermodynamic equilibrium where the linear phenomenological relations (3.81) between the forces F_α and the fluxes I_α and thus also for their variations are valid:

$$\delta I_\alpha = \sum_\beta L_{\alpha\beta} \delta F_\beta \quad . \qquad (7.43)$$

Inserting (7.43) into (7.42) we obtain

$$\frac{dL}{dt} = - \sum_{\alpha,\beta} L_{\alpha\beta} \delta F_{\alpha} \delta F_{\beta} \quad . \tag{7.44}$$

Now, the condition $dL/dt \leq 0$ as required for the Liapunov-stability of the equilibrium is reduced to the condition that the matrix $L_{\alpha\beta}$ of the linear phenomenological coefficients is positive definite. This latter property, however, is a direct consequence of the second law of thermodynamics as we have shown in (3.82). With this conclusion we have reconfirmed our preliminary result of the preceding section in a formally precise way: a sufficient condition for the stability of the equilibrium is a) a positive-definite capacitance matrix such that $L \geq 0$ and b) the second law of thermodynamics such that $dL/dt \leq 0$. Let us emphasize once more the significance of the equivalence between $dL/dt \leq 0$ and the second law in the form of (3.82). This equivalence, however, is valid only in the range of validity of the linear relations in (3.81). If the fluxes I_{α} were some nonlinear functions of the forces F_{α} as will be the case in situations far from the thermodynamic equilibrium, $dL/dt \leq 0$ is no longer guaranteed by the second law and possibly may no longer be valid at all.

To conclude this section let us remark that Onsager's reciprocity relations $L_{\alpha\beta} = L_{\beta\alpha}$, (3.83), seem to be completely irrelevant for the stability property of the equilibrium. This result is surprising from a physical point of view since $L_{\alpha\beta} = L_{\beta\alpha}$ is considered as a fundamental property of the equilibrium state and related to the reversibility of the microscopic motion of the system. As a matter of fact, we did not make use of $L_{\alpha\beta} = L_{\beta\alpha}$ but the extreme opposite, namely an antisymmetric matrix $L_{\alpha\beta} = -L_{\beta\alpha}$ would lead to $dL/dt = 0$ in (7.44). This means that only a very weak version of the reciprocity property is needed: the matrix $L_{\alpha\beta}$ must have a nonvanishing symmetric part $L^{s}_{\alpha\beta} = (L_{\alpha\beta} + L_{\beta\alpha})/2 \neq 0$.

7.4 Glansdorff-Prigogine Criterion for the Stability of Nonequilibrium Steady States

As we have emphasized in the preceding section, the stability of equilibrium states crucially depends on the validity of $dL/dt \leq 0$ which is a direct consequence of the second law of thermodynamics only within the range of the linear relationships between the fluxes I_{α} and the forces F_{α}. Since in general this linearity will not be valid in the vicinity of steady states arbitrarily far from equilibrium, we cannot transfer the above stability proof to such states. In this situation, we invert the order of arguments: whereas for the equilibrium $dL/dt \leq 0$ is satisfied due to the second law and $L \geq 0$, i.e., the positive definiteness of the capacitance matrix, appears as the condition or criterion for stability, we now assume $L \geq 0$ to be guaranteed by the structure of the system and take $dL/dt \leq 0$ as the condition or criterion for stability which then has to be evaluated separately for each steady state of a particular system. According to (7.42) the criterion $dL/dt \leq 0$ can be expressed as

$$\sum_{\alpha} \delta F_{\alpha} \cdot \delta I_{\alpha} \geq 0 \tag{7.45}$$

where the δF_{α} and δI_{α} are now variations around the steady state under consideration. Eq. (7.45) as a differential or local stability criterion of nonequilibrium states has been introduced by GLANSDORFF and PRIGOGINE (1971). The expression on the left-hand side of (7.45) is called the excess entropy production to be distinguished from the total entropy production (3.78). Comparing (7.45) and (3.78) we see that the excess entropy production is the second-order variation $\delta^2 P$ of the total entropy production $P = \dot{S}$ as given by (3.78).

It should be emphasized that the assumption $L \geq 0$ is far from being trivial or unproblematic but implies a positive capacitance matrix for all possible steady states or, in other words, stable equilibrium states under any external equilibrium conditions for the system. By the expression "equilibrium conditions" we mean external conditions under which the system is capable of realizing an equilibrium state. The assumption $L \geq 0$ therefore excludes for example any kind of phase transitions.

In context with the examples for instable networks in Chapter 6 it is elucidating to look for candidates of processes giving negative contributions to the excess entropy production of (7.45) and thus possibly leading the system into an instability. As an example let us consider an autocatalytic reaction $\nu X + Y \rightleftharpoons (\nu + 1)X$ as discussed in Section 6.1. With

$$J = kX^{\nu}Y - k'X^{\nu + 1} \tag{7.46}$$

for the flux of the reaction we obtain for its affinity by comparing (7.46) with (4.23)

$$A = A^f - A^r = RT \ln \frac{kX^{\nu}Y}{k'X^{\nu + 1}} = RT \ln \frac{kY}{k'X} \quad . \tag{7.47}$$

Due to (3.79) the force of the reaction is given as $F = A/T$. Expressing now the variations of X and Y by the molar extent of the reaction, $\delta X = \delta\xi$, $\delta Y = -\delta\xi$ we easily derive

$$\left.\begin{aligned} \delta J &= \left[\nu k X^{\nu-1} Y - (k + (\nu+1)k')X^{\nu}\right]\delta\xi \\ \delta F &= -R\left(\frac{1}{X} + \frac{1}{Y}\right)\delta\xi \end{aligned}\right\} \quad . \tag{7.48}$$

Whereas for $\delta\xi > 0$ we always have $\delta F < 0$, the expression for δJ may become positive particularly for small X. Thus, the product $\delta F \cdot \delta J$ may give negative contributions to the left-hand side of (7.45). Whether this negative contribution actually leads to an instability clearly depends on the total balance of the excess entropy production. On the other hand, the reader may easily prove that ordinary chemical reactions always have a positive excess entropy production as separate contributions to the left-hand side of (7.45), cf. problem 1.

Frequently the stability criterion of Glandorff and Prigogine is expressed as an evolution criterion. By interpreting the variations $\delta\mu_i$ and δc_i in (7.24) as variations during a real time evolution of the system we can formulate the assumption of a positive definite capacitance matrix equivalently as

$$\sum_C V_i \frac{d\mu_i}{dt}\frac{dc_i}{dt} = \sum_C V_i \frac{d\mu_i}{dt} \cdot J_i \geq 0 \tag{7.49}$$

where the symbol C reminds us to sum over all ports i of the capacitances. Applying once more Tellegen's theorem to (7.49) and recalling the definition of the generalized forces and fluxes, cf. (7.39), (7.41), and (7.42), we obtain from (7.49)

$$\frac{d_F P}{dt} = \sum_\alpha \frac{dF_\alpha}{dt} \cdot I_\alpha \leq 0 \quad . \tag{7.50}$$

The notation on the left-hand side of (7.50) indicates that (7.50) is that part of the time derivative of the total entropy production

$$P = \dot{S} = \sum_\alpha F_\alpha \cdot I_\alpha \tag{7.51}$$

which is associated with the time change of the forces F_α:

$$\frac{dP}{dt} = \frac{d_F P}{dt} + \frac{d_I P}{dt} \tag{7.52}$$

$$\frac{d_F P}{dt} = \sum_\alpha \frac{dF_\alpha}{dt} \cdot I_\alpha \quad , \qquad \frac{d_I P}{dt} = \sum_\alpha F_\alpha \frac{dI_\alpha}{dt} \quad . \tag{7.53}$$

Within the range of validity of the linear relations (3.81) the reader easily verifies by making use of Onsager's reciprocity relations (3.83)

$$\frac{d_I P}{dt} = \frac{d_F P}{dt} \quad \text{and} \quad \frac{d_F P}{dt} = \frac{1}{2}\frac{dP}{dt} \tag{7.54}$$

such that (7.50) once more proves the theorem of minimal entropy production within the range of linear irreversible thermodynamics, cf. (3.84). For situations far from thermodynamic equilibrium this conclusion is no longer valid.

Returning now to the general case, (7.50) can be considered as a generalized evolution criterion for all real processes. This criterion includes the principle of minimum entropy production in the linear range. An evolution criterion, however, can immediately be retranslated into a stability criterion: if for all variations δF_α in the vicinity of a steady state the condition (7.50) is violated, i.e., if

$$\sum_\alpha \delta F_\alpha \cdot I_\alpha \geq 0 \tag{7.55}$$

the steady state is stable. Indeed, (7.55) is equivalent to the condition of positive excess entropy production (7.45). This is easily seen by writing $I_\alpha = \bar{I}_\alpha + \delta I_\alpha$ where $\bar{I}_\alpha$ is the steady state value of flux I_α and by making use of the relation

$$\sum_\alpha \delta F_\alpha \cdot \bar{I}_\alpha = 0 \quad . \tag{7.56}$$

Eq. (7.56) is derived from the very definition of a steady state, namely $\bar{J}_i = 0$ along all capacitance ports i such that together with Tellegen's theorem we have

$$-\sum_C V_i \delta\mu_i \bar{J}_i = \sum_R V_i \delta\mu_i \bar{J}_i = \sum_\alpha \delta F_\alpha \cdot \bar{I}\alpha = 0 \tag{7.57}$$

cf. (7.35).

7.5 Uniqueness of Steady States

The stability criteria which we have discussed in the preceding sections of this chapter are local or differential criteria since they are formulated in terms of second-order variations and their time derivatives. On the other hand, we have seen in Chapter 6 that it would be very helpful to gain knowledge about the global behaviour of a network for classifying its properties and deciding whether it may serve as an adequate model for a certain biological phenomenon or not. A consequent network theory of stability should therefore aspire to formulate topological criteria at least for the principal types of global behaviour, in particular for the existence of unique or multiple steady states and for deciding whether the state of the system tends into a steady state at $t \longrightarrow \infty$ or possibly performs periodic or nonperiodic bounded or unbounded motions. The present state of the art of this kind of network theory is still very far from making such general predictions. Only some empirical rules like that in Chapter 6 concerning the implications of nonlinear closed loops with positive or negative feedbacks and a few sufficient conditions for excluding certain types of instabilities can be given.

An example of the latter kind of condition will be presented in this section. It says that a network consisting of capacitances, 2-port elements, 0- and 1-junctions as introduced in Chapter 4 always has a unique steady state if no closed loops appear in the network. It is clear that as a closed loop we understand any series of elements connected by bonds which lead back into the initial element. A transducer, although introduced in Section 4.5 as a shortwriting for a closed loop consisting of a 0- and a 1-junction, will not interfere with the proof to be given for the above uniqueness statement. Examples for closed loops may be found in the networks for an enzyme-catalyzed reaction or pore transport (4.38) or (5.1), for carrier transport (5.41), for active transport (5.50) and for an autocatalytic reaction (6.3). Except for the latter one, all above-quoted examples showed a unique steady state despite the appearance of a closed loop. This means that the absence of closed loop in a network can only be a sufficient condition for a unique steady state and is certainly far from being necessary. As regarding the 2-port elements for reactions and diffusion, we shall assume that the rates of the reaction and diffusion fluxes satisfy the product ansatz in terms of the concentrations. After having read the proof to be given below the reader may verify that a less incisive assumption as, for example, rates which are monotonically increasing functions of the concentrations, would also be sufficient.

To prove the above formulated uniqueness theorem, we consider an arbitrary but loop-free network N and assume that it has a unique steady state for arbitrary but fixed values of its external variables $A_1, \ldots A_n$. As mentioned earlier, an external variable can be visualized as an infinitely large capacitance. Now we convince ourselves that the uniqueness remains valid if one of the external variables, say A_1, is replaced by a 0-junction to which an internal variable X and an additional reaction 2-port element R are connected. The new element R may possibly involve further internal or external variables Y and Z, but its connection to N is assumed to generate no closed loop together with the elements of N:

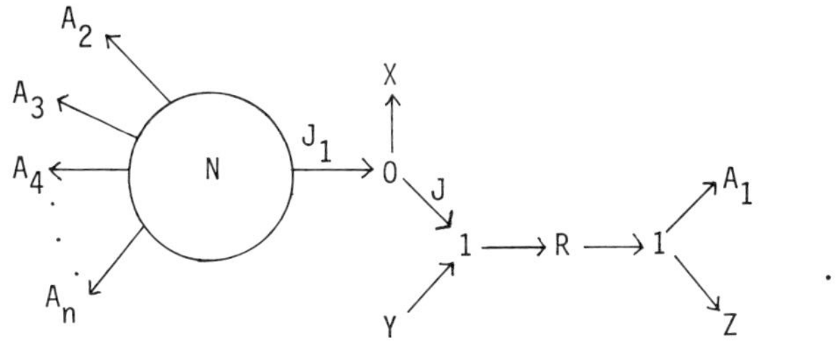

(7.58)

The flux J_1 along the bond from N to X has the structure

$$J_1 = J_1^f - k_1'X \tag{7.59}$$

where according to the absence of closed loops the forward rate J_1^f is independent of X. Let us consider now the steady state of N together with the steady state flux $\bar{J}_1$ as a function of X at fixed values of A_2, ... A_n. Since $\bar{J}_1$ for given values of X is a steady state property of N and the steady state of N had been assumed to be unique, we may consider $\bar{J}_1$ as a unique function of X: $\bar{J}_1 = \bar{J}_1(X)$. From (7.59) we also conclude that

$$\bar{J}_1(X=0) > 0 \quad , \quad \lim_{X \to +\infty} \bar{J}_1(X) = -\infty \quad , \quad \frac{d\bar{J}_1(X)}{dX} < 0 \quad . \tag{7.60}$$

The flux J of the additional reaction R has the structure

$$J = kXY - k'A_1Z \quad . \tag{7.61}$$

If at least one of the variables Y and Z is an internal variable, the steady state value of J vanishes, $\bar{J} = 0$, and hence also $\bar{J}_1 = 0$. According to (7.60) the condition $\bar{J}_1(X) = 0$ has a unique solution $X = \bar{X}_o$ such that (7.61) for $\bar{J} = 0$ becomes

$$\bar{J} = k\bar{X}_o\bar{Y} - k_1'A_1\bar{Z} = 0 \quad . \tag{7.62}$$

It both Y and Z are internal variables we have

$$\frac{dY}{dt} = -J \quad , \quad \frac{dZ}{dt} = J \tag{7.63}$$

such that Y + Z = C = const. Eq. (7.62) together with $\bar{Y} + \bar{Z} = C$ has a unique solution for $\bar{Y}$ and $\bar{Z}$ for any given values of $\bar{X}_o > 0$.

If only one of the variables Y and Z, say Y, is internal whereas $Z = Z_0$ is fixed and given, (7.62) is reduced to

$$k\overline{X}_0\overline{Y} - k'A_1Z_0 = 0 \tag{7.64}$$

which again has a unique solution for $\overline{Y}$ for any given value of $\overline{X}_0 > 0$ and $Z_0 > 0$.

Finally we have the possibility that both Y and Z are fixed and given external variables: $Y = Y_0$, $Z = Z_0$. Now the steady-state condition reads

$$\overline{J}_1(X) = \overline{J}(X) \tag{7.65}$$

where

$$\overline{J}(X) = kXY_0 - k'A_1Z_0 \quad . \tag{7.66}$$

From (7.66) we immediately have

$$\overline{J}(X = 0) < 0 \quad , \quad \lim_{X \to +\infty} \overline{J}(X) = +\infty \quad , \quad \frac{d\overline{J}(X)}{dX} > 0 \tag{7.67}$$

which together with (7.60) again guarantees a unique solution $X = \overline{X}$ of (7.65).

Our proof is completed by remarking that any loop-free network can be generated by a series of extension steps as described above. Since each single step does not affect the uniqueness property, the resulting network again has a unique steady state. Let us once more stress the significance of the assumption that the network is free of closed loops for our proof. If, for example, the reaction 2-port R in (7.58) does generate a closed loop together with some of the bonds and elements of N, the forward rate J_1^f of J_1 could possibly become X-dependent since the generated loop poses additional internal KCL constraints between the steady state fluxes. As a consequence, we could no longer take $d\overline{J}_1/dX < 0$ in (7.60) as guaranteed although this condition might be satisfied despite the presence of the closed loop. If, however, there really is a nonmonotonic dependence of $\overline{J}_1$ on X, the steady state conditions $\overline{J}_1 = 0$ or $\overline{J}_1 = \overline{J}$ could have two or more solution as in the case for example for the autocatalytic reaction system in Section 6.1, cf. Fig. 10.

7.6 Globally and Asymptotically Stable Networks

If a network has been shown to have a unique steady state, either by application of the uniqueness criterion of Section 7.5 or by an individual inspection, there are still three possibilities for its global behaviour: a) the state of the network tends into the steady state as $t \longrightarrow \infty$, i.e., global and asymptotic stability of the steady state; b) the state of the network performs bounded periodic or nonperiodic motions which do not lead into the steady state; c) at least one of the variables of the networks tends to infinite values at finite or infinite times. Again, it would be very helpful to have network topological criteria for distinguishing these three types of global behaviour; but such a general prediction is again still beyond the present state of the theory. What actually can be predicted about the global behaviour of a system on the basis of network topological properties is first of all that the case c) among the above listed alternatives can be excluded at least for networks with elements as introduced in Chapter 4.

We shall not try to give a rigorous proof of this statement but make it plausible at least regarding the essential line of the argument. First of all we convince ourselves that a single concentration variable of a capacitance in the network, say X_1, cannot tend to infinite values if all other variables of the network remain finite. This is immediately concluded from the kinetic equation for X_1,

$$\frac{dX_1}{dt} = \sum_i J_{1i} \equiv \Phi_1(X_1, X_2, \ldots) \tag{7.68}$$

where J_{1i} are the fluxes of reaction or diffusion 2-ports which are connected to the capacity port of X_1. Assuming the usual product ansatz for the fluxes J_{1i} we see that increasing values of X_1 will eventually cause the reverse rates of J_{1i}, i.e., the rates directed from X_1 to other capacities, to exceed the forward rates of J_{1i} such that dX_1/dt becomes negative. This argument includes the case of autocatalytic reactions: increasing values of X_1 in the reaction $A + \nu X_1 \rightleftharpoons (\nu + 1)X_1$ will eventually lead to negative contributions to dX_1/dt.

What remains to be excluded is the case that pairs, triples, ... etc. of variables which are directly connected by reaction or diffusion fluxes tend to infinite values simultaneously. Let X_1, X_2 be such a pair and let the flux which couples X_1 to X_2 be a reaction flux

$$J = kX_1^{\nu_1}X_2^{\nu_2} \ldots - k'X_1^{\nu_1'}X_2^{\nu_2'} \ldots . \tag{7.69}$$

where further variables X_3, X_4, ... may be involved with bounded values. In (7.69) we have assumed the rates of the fluxes to satisfy a product ansatz. Again, a less incisive assumption as, for example, rates which are monotonically increasing functions of the concentrations would be sufficient. The contributions of J to the time changes of X_1 and X_2 are

$$\frac{dX_1}{dt} = -(\nu_1 - \nu_1')J + \ldots \quad , \qquad \frac{dX_2}{dt} = -(\nu_2 - \nu_2')J + \ldots \quad . \tag{7.70}$$

Without loss of generality we assume that J diverges to positive infinite values if $X_1 \longrightarrow \infty$ and $X_2 \longrightarrow \infty$ (otherwise we interchange the roles of k, ν_1, ν_2 and k', ν_1', ν_2'). This implies that at least one of the inequalities $\nu_1 - \nu_1' > 0$, $\nu_2 - \nu_2' > 0$ is satisfied which in turn causes either dX_1/dt or dX_2/dt or both in (7.70) to become negative for increasing values of X_1, X_2 such that at least one of the variables X_1, X_2 is bounded. This latter consequence is in contradiction to the previous assumption of simultaneously diverging variables X_1, X_2. The argument is readily extended to the case of simultaneously diverging variables $X_1, X_2, \ldots X_n$ which are pairwise coupled by fluxes.

Having excluded the possibility of diverging concentration variables in the network, one would now wish to have a criterion for distinguishing between the remaining possibilities a) of a globally and asymptotically stable state and b) of bounded pariodic or nonperiodic motions around the steady state. In Section 6.4 we have already discussed a local criterion for a limit cycle in 2-variable systems. This criterion interprets an instability of the steady state as a possible onset of a limit cycle. Clearly, it is therefore not a complete global criterion and does not globally ensure the existence of a limit-cycle solution. The reader who is interested in rigorous mathematical criteria for limit cycles is referred to the monograph of MINORSKY (1974). So far the known mathematical criteria have not yet been translated into the network language.

What actually can be derived on the basis of a network representation is a negative criterion for limit cycles which says that loop-free networks not only have a unique steady state as argued in Section 7.5 but that this steady state also is globally and asymptotically stable. This statement evidently excludes the possibility of a limit cycle. The proof of this criterion has an analytic and a network topological part. In the analytic part one shows that the system of differential equations

$$\frac{dX_i}{dt} = \Phi_i(X_1, \ldots X_n) \qquad 1 \leq i \leq n \tag{7.71}$$

is globally and asymptotically stable if the following conditions are satisfied:

1) the system has a unique steady state $\overline{X}_1, \overline{X}_2, \dots \overline{X}_n$,

$$\Phi_i(\overline{X}_1, \overline{X}_2, \dots \overline{X}_n) = 0 \qquad 1 \leq i \leq n \tag{7.72}$$

with positive components $\overline{X}_i > 0$;

2) Φ_i is a monotonically decreasing function of X_i,

$$\Phi_{ii} \equiv \frac{\partial \Phi_i}{\partial X_i} < 0 \tag{7.73}$$

and

$$\lim_{X_i \to 0} \Phi_i > 0 \quad , \qquad \lim_{X_i \to \infty} \Phi_i = -\infty \tag{7.74}$$

at arbitrarily fixed values of X_j, $j \neq i$;

3) either

$$\Phi_{ij} \equiv \frac{\partial \Phi_i}{\partial X_j} = 0 \quad \text{and} \quad \Phi_{ji} \equiv \frac{\partial \Phi_j}{\partial X_i} = 0 \tag{7.75}$$

or

$$\Phi_{ij} \cdot \Phi_{ji} > 0 \tag{7.76}$$

at all points of the phase space $X_1, X_2, \dots X_n$. Eq. (7.76) implies that if Φ_i depends on X_j then Φ_j also depends on X_i, Φ_{ij} and Φ_{ji} always have the same sign and neither Φ_{ij} nor Φ_{ji} can change its sign.

Again, we shall not give a detailed proof of the above theorem. Let us just mention that the proof relies on the fact that under the above conditions one can construct a series of n-dimensional cubes

$$C_\nu : a_i^{(\nu)} \leq X_i \leq b_i^{(\nu)} \quad , \qquad 1 \leq i \leq n \quad , \qquad \nu = 1,2, \dots \tag{7.77}$$

such that all C_ν's contain the steady state, every C_ν contains its successor $C_{\nu+1}$ and the directional field $(\Phi_1, \Phi_2, \dots \Phi_n)$ of the differential equation (7.71) on the surface of a C_ν is always directed into the interior of C_ν. It is clear that the trajectory of the system, i.e., its state expressed by the variables $X_1(t), X_2(t), \dots X_n(t)$ as functions of time, can then never leave a cube C_ν but approaches the steady state as $t \longrightarrow \infty$ irrespective of the initial state at some time t_o.

In the network topological part of our proof we now convince ourselves that a loop-free network indeed satisfies the above listed conditions 1), 2) and 3). After our considerations in the preceding section and at the beginning of this section, this point almost needs no further argument. Condition 1) has been shown already in Section 7.5), condition 2) has essentially been utilized when proving the uniqueness property of Section 7.5 and condition 3) is a consequence of the fact that two capacities X_1, X_2 which are coupled by a flux of a reaction 2-port in a loop-free network are either on the same side or on opposite sides of this 2-port such that either $\Phi_{12} < 0$ and $\Phi_{21} < 0$ or $\Phi_{12} > 0$ and $\Phi_{21} > 0$.

The analytic part of the above proof is evidently not restricted to loop-free networks but may be applied as a criterion of global and asymptotical stability to arbitrary networks as an individual test for the network with respect to conditions 1), 2) and 3).

7.7 Global Stability Techniques

For an evaluation of the global behaviour of a network involving closed loops an individual investigation of its equations of motions will be inevitable. The equations of motion of a network are systems of first-order differential equations for the capacitive variables and are to be derived from the structure of the network. There exist a few general techniques which have turned out to be helpful when investigating the global behaviour of systems of first-order differential equations. For a general survey the reader is referred to the monograph of MINORSKY (1974). We shall briefly describe two rather simple techniques of such a kind in this section.

The first one of these techniques may be called a geometrical phase space analysis and has been utilized already in section 6.3, Fig. 12. Let us consider the case of a network with two capacitive variables X_1 and X_2 and let the equations of motion be

$$\frac{dX_i}{dt} = \Phi_i(X_1, X_2) \qquad i = 1,2 \quad . \tag{7.78}$$

For reaction and diffusion fluxes satisfying the usual product ansatz for their rates the $\Phi_i(X_1, X_2)$ will be polynomials in X_1 and X_2 such that the equations $\Phi_i(X_1, X_2) = 0$ for $i = 1,2$ will define two regular curves W_i, $i = 1,2$, in the X_1-X_2-plane, the so-called phase space. Evidently, the intersection points of W_1 and W_2 are the steady states of the network. Moreover, the curves W_1 and W_2 will split up the full phase space into bounded and unbounded parts P_1, P_2 .... Within each of the P_α the components Φ_i of the directional field (Φ_1, Φ_2) of $dX_i/dt = \Phi_i$

have definite signs such that the vector (Φ_1, Φ_2) has a definite direction with respect to the coordinate axes of X_1 and X_2. Marking now these directions into the P_α gives at least a qualitative impression of what will happen with the state of the network. As Fig. 12 shows, stable steady states as point C in Fig. 12(a) can unambiguously be distinguished from steady states with saddle-point character as point C in Fig. 12(c) or from those which are instable with respect to all variations as point A in Fig. 12(a) - (c).

The weak points of this technique are first of all the difficulties with the geometrical perception of directional fields $(\Phi_1, \Phi_2, \ldots \Phi_n)$ in phase spaces of more than two dimensions, i.e., for networks with more than two variables. In such cases, projections of the full n-dimensional directional field $(\Phi_1, \Phi_2, \ldots \Phi_n)$ into 2-dimensional coordinate planes (X_i, X_j) will at least lead to a classification of the stability character of steady states. Actually, such a projection procedure is the basis for proving the analytic part of the global stability theorem for loop-free networks in the preceding section.

A second weak point of the geometrical phase space analysis is the fact that it will be very difficult if not impossible to detect the onset of a limit cycle at a critical or bifurcation point as defined in Section 6.4 for two variables by

$$\lambda^c = -\frac{1}{2}(\Phi_{11}^c + \Phi_{22}^c) = 0 \tag{7.79}$$

just by looking at the signs of the directional field in the phase space. In problem 7 the reader may convince himself that the geometrical phase space analysis makes the tendency towards a limit cycle behaviour very suggestive, however, whether or under which conditions it really occurs finally remains to be decided by an analytic criterion like that in (7.79).

The second technique to be briefly described in this section is that of global Liapunov-functions. Indeed, we have made use of the Liapunov-technique already in Sections 7.2 and 7.4 when defining a Liapunov-function L by

$$L = -\frac{1}{2}\,\delta^2 S \tag{7.80}$$

where $\delta^2 S$ is the second-order variation of the network entropy either around the equilibrium or some other nonequilibrium steady state. If now $L \geq 0$ and $L = 0$ only in the steady state (equilibrium or not) and if one can show that the equations of motion of the system imply $dL/dt \leq 0$ such that $dL/dt = 0$ only in the steady state, the steady state is stable. This stability statement, however, is restricted to a differential vicinity of the steady state since L in (7.80) is defined as a differential variation and the derivation of $dL/dt \leq 0$ will thus necessarily include a linearization of the equations of motion except for the case that the equations of motion are a priori linear.

On the other hand, if one succeeds in formulating a global Liapunov function $L \geq 0$ and proving that $dL/dt \leq 0$ holds all over the complete phase space, the stability statement resulting from that evidently implies the existence of a unique steady state which is globally and asymptotically stable. The uniqueness simply follows from the impossibility of having two different steady states which both are globally stable. As an example of such a situation, let us consider a network consisting of capacitances and reaction or diffusion 2-ports which are arbitrarily connected only by 0-junction whereas 1-junctions are not present in the network. Note that nevertheless the network may involve now an arbitrary number of closed loops. Let us describe the network in the concentration-flux language such that all capacitances are 1-ports and let X_1, X_2, ... be the concentration variables of the capacitances. Because of the absence of 1-junctions the equations of motion of the network are linear. Let the equations of motion be given as

$$\frac{dX_i}{dt} = \sum_j J_{ij} + \sum_e J_{ie} \qquad (7.81)$$

where J_{ij} is the reaction or diffusion flux between two capacitances X_j and X_i with reference orientation from j to i, which upon assuming the product ansatz for the rates to be satisfied reads

$$J_{ij} = k_{ij}X_j - k_{ji}X_i \qquad (7.82)$$

and J_{ie} is the reaction or diffusion flux from some external bath at constant concentration A_e to X_i:

$$J_{ie} = k_{ie}A_e - k_{ei}X_i \quad . \qquad (7.83)$$

Summing (7.81) over all internal capacitances X_i and making use of the antisymmetry of J_{ij}, $J_{ji} = -J_{ij}$, we readily obtain

$$\sum_i \frac{dX_i}{dt} = \sum_{i,e} J_{ie} \quad . \qquad (7.84)$$

From (7.81) and (7.84) we derive two important relations for the fluxes $\bar{J}_{ij}$, $\bar{J}_{ie}$ in the steady state ($dX_i/dt = 0$) namely

$$\sum_j \bar{J}_{ij} + \sum_e \bar{J}_{ie} = 0 \quad , \qquad \sum_{i,e} \bar{J}_{ie} = 0 \tag{7.85}$$

which will be needed below for the stability proof.

Next we define a Liapunov function L by

$$L = \sum_i \lambda(X_i, \bar{X}_i) \tag{7.86}$$

where

$$\lambda(X_i, \bar{X}_i) = X_i \ln \frac{X_i}{\bar{X}_i} - X_i + \bar{X}_i \tag{7.87}$$

and $\bar{X}_i$ is an arbitrary steady state of (7.81). From (7.87) we easily derive

$$\lambda(\bar{X}_i, \bar{X}_i) = 0 \tag{7.88}$$

$$\frac{d\lambda(X_i,\bar{X}_i)}{dX_i} = \ln \frac{X_i}{\bar{X}_i} = \begin{cases} < 0 & X_i < \bar{X}_i \\ > 0 & X_i > \bar{X}_i \end{cases} \tag{7.89}$$

$$\frac{d^2\lambda(X_i,\bar{X}_i)}{dX_i^2} = \frac{1}{X_i} > 0 \quad . \tag{7.90}$$

Eqs. (7.89) and (7.90) say that $\lambda(X_i, \bar{X}_i)$ is monotonically decreasing for $X_i < \bar{X}_i$ and monotonically increasing for $X_i > \bar{X}_i$, cf. Fig. 13. Evidently, this behaviour of $\lambda(X_i, \bar{X}_i)$ implies that if we could prove $dL/dt \leq 0$ with the equality sign being valid only for $X_i = \bar{X}_i$ the network would tend into the steady state $\bar{X}_i$ as $t \longrightarrow \infty$ wherever it starts.

Calculating now dL/dt from (7.86) and (7.87) we obtain upon inserting (7.81), (7.82) and (7.83)

$$\begin{aligned} \frac{dL}{dt} &= \sum_i \frac{dX_i}{dt} \cdot \ln \frac{X_i}{\bar{X}_i} = \\ &= \sum_{i,j} (k_{ij}X_j - k_{ji}X_i) \ln \frac{X_i}{\bar{X}_i} + \\ &\quad + \sum_{i,e} (k_{ie}A_e - k_{ei}X_i) \ln \frac{X_i}{\bar{X}_i} = \\ &= \sum_{i,j} k_{ij}X_j \ln \frac{X_i\bar{X}_j}{\bar{X}_iX_j} + \sum_{i,e} k_{ie}A_e \ln \frac{X_i}{\bar{X}_i} + \sum_{i,e} k_{ei}X_i \ln \frac{\bar{X}_i}{X_i} \end{aligned} \tag{7.91}$$

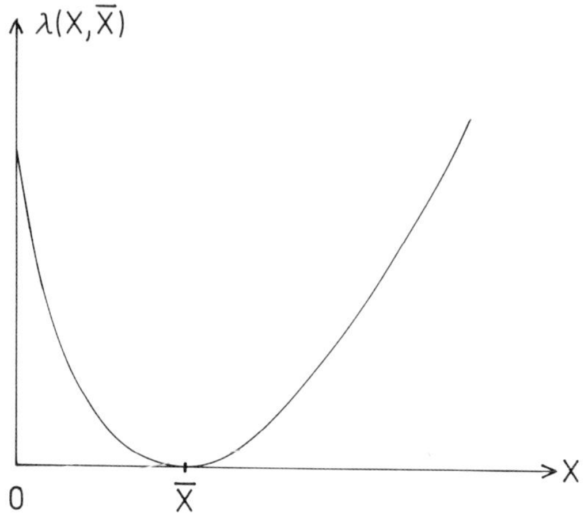

Fig. 13. $\lambda(X, \bar{X}) = X \ln (X/\bar{X}) - X + \bar{X}$ as a function of X

where in the last but one step we have interchanged the summation variables i,j in the reverse rate of J_{ij}. Making use now of the inequality $\ln x \leq x - 1$ for all $x > 0$ the equality sign being valid only for $x = 1$, and noticing $k_{ij}X_j > 0$, $k_{ie}A_e > 0$, $k_{ei}X_i > 0$ we derive the following inequality from (7.91):

$$
\begin{aligned}
\frac{dL}{dt} &\leq \sum_{i,j} k_{ij}\left(\frac{X_i}{\bar{X}_i}\bar{X}_j - X_j\right) + \sum_{i,e} k_{ie}A_e\left(\frac{X_i}{\bar{X}_i} - 1\right) + \sum_{i,e} k_{ei}\,(\bar{X}_i - X_i) \\
&= \sum_i \frac{X_i}{\bar{X}_i}\left[\sum_j (k_{ij}\bar{X}_j - k_{ji}\bar{X}_i) + \sum_e (k_{ie}A_e - k_{ei}\bar{X}_i)\right] \\
&\qquad - \sum_{i,e}(k_{ie}A_e - k_{ei}\bar{X}_i) \\
&= \sum_i \frac{X_i}{\bar{X}_i}\left(\sum_j \bar{J}_{ij} + \sum_e \bar{J}_{ie}\right) - \sum_{i,e}\bar{J}_{ie} = 0
\end{aligned}
\tag{7.92}
$$

where another interchange of summation variables has been performed and use has been made of the steady state relations (7.85). Note that in (7.92) the equality sign is valid only if all capacitances X_i have their steady state values $\bar{X}_i$. With (7.92) we have completed our proof that if there exists a steady state $\bar{X}_i$ of the network, it is globally and asymptotically stable and thus also unique. What remains to be proved is that there exists at least one steady state $\bar{X}_i$ of (7.81). This can readily be shown by an stepwise extension procedure very similar as that described in Section 7.5, the only difference being the fact that for the proof of the existence of at least one steady state we do not even require the uniqueness of the construction in Section 7.5 and thus do not have to assume that the network is free of closed loops.

The above proof is readily extended to include the appearance of 1-junctions of the type

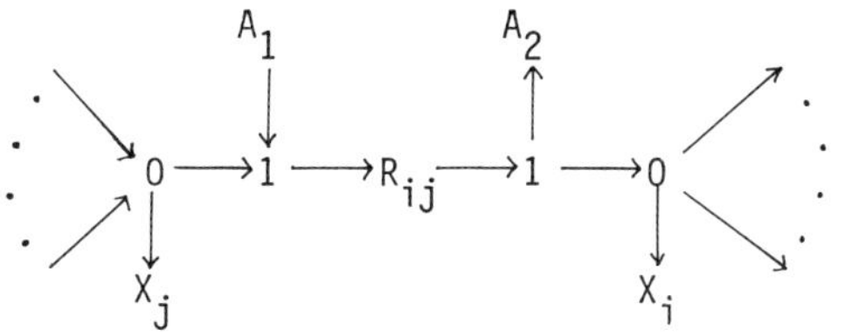

(7.93)

in the network where A_1 and A_2 are fixed external concentrations. Eq. (7.93) compared to the situation where A_1 and A_2 are absent is equivalent to a redefinition of the rate constants k_{ij}, k_{ji}. With this generalization our stability proof now includes the network models for pore transport in (5.1), for carrier transport in (5.41) and for active transport in (5.50).

Problems

1) Prove that a single chemical reaction of the type $X_1 + X_2 + \ldots \rightleftharpoons Y_1 + Y_2 + \ldots$ where none of the reactants simultaneously appears on both sides of the reaction, always has a positive excess entropy production.
2) Argue how it could in principle be possible to have a total negative excess entropy production in a network which consists of capacitances and chemical reactions of the type as assumed in problem 1 (Think of closed loops).
3) Apply conditions 1), 2) and 3) as expressed by (7.72) to (7.76) as a criterion for global and asymptotical stability to the networks throughout this book and show that those networks which have already been identified as globally and asymptotically stable indeed satisfy the conditions whereas those showing multiple steady states or limit cycles violate conditions 2) or 3) or even both.
4) Prove that the quantity H of (2.51) which is conserved in the original version of the Volterra-Lotka model can be expressed by the global Liapunov function as defined in (7.86), (7.87). (Interchange the arguments X_i and $\overline{X}_i$ in (7.87).).
5) Apply the global Liapunov function L as defined in (7.86), (7.87) to determine the global behaviour of the Ising model of Section 2.5 and of the autocatalytic reaction networks as described in (6.3).
6) Calculate the lowest nonvanishing order ΔL of L of (7.86), (7.87) in terms of the variations $\delta X_i = X_i - \overline{X}_i$ around the steady state $\overline{X}_i$ and its time derivative $d(\Delta L)/dt$. Prove that the stability condition $d(\Delta L)/dt \leq 0$ now coincides with the

differential stability condition (7.45) for the excess entropy production of the reaction and diffusion 2-ports as to be calculated from the fluxes (7.82) and (7.83) and its corresponding affinities

$$A_{ij} = RT \ln \frac{k_{ij}X_j}{k_{ji}X_i} \quad , \qquad A_{ie} = RT \ln \frac{k_{ie}A_e}{k_{ei}X_i}$$

cf. (7.47).

7) Perform the phase space analysis as described in Section 7.7 for the network (6.27). Make plausible that the directional field in this case may possibly give rise to a limit cycle oscillation. Is it possible to determine the bifurcation point form the phase space analysis?

References

Adam, G.: Z. Naturforsch. 23 b, 181 (1968)
Adam, G.: In: Snell, F., Wolken, J., Iverson, G.J., Lam, J. (Eds.): *Physical Principles of Biological Membranes. Gordon and Breach Science*. New York: 1970.
Balslev, I., Degn, H.: J. Theor. Biol. 49, 173 (1975).
Bass, L., Moore, W.J.: J. Membrane Biol. 12, 361 (1973).
Bebber, H.J.: Diploma Thesis. Aachen 1975.
Bernstein, J.: Pflügers Arch. ges. Physiol. 92, 521 (1902).
Bernstein, J.: *Elektrobiologie*. Braunschweig: Vieweg 1912.
Blumenthal, R., Changeux, J.P., Lefever, R.: J. Membrane Biol. 2, 351 (1970).
Callen, H.B.: *Thermodynamics*. New York, London, Sydney: Wiley 1960.
Chapman, D., Leslie, R.B.: *Molecular Biophysics*. Contemporary Science Paperbacks. Edingburgh and London: Oliver and Boyd 1967.
Cole, K.S.: *Membranes, Ions and Impulses*. Berkeley and Los Angeles: University of California Press 1968.
Danielli, J.F., Davson, H.: J. cell. comp. Physiol. 5, 495 (1935).
De Groot, S.R.: *Thermodynamics of Irreversible Processes*. Amsterdam: North Holland Publishing 1951.
De Groot, S.R., Mazur, P.: *Nonequilibrium Thermodynamics*. Amsterdam: North Holland Publishing 1962.
Forst, W.: *Theory of Unimolecular Reactions*. New York, London: Academic Press 1973.
Glansdorff, P., Prigogine, I.: *Thermodynamic Theory of Structure, Stability and Fluctuations*. London, New York, Sydney, Toronto: Wiley-Interscience 1971.
Goel, N.S., Maitra, S.C., Montroll, E.W.: *On the Volterra and other Nonlinear Models of Interacting Populations*. New York, London: Academic Press 1971.
Gotoh, H.: J. theor. Biol. 53, 309 (1975)
Griffith, J.S.: J. theor. Biol. 20, 202 (1968)
Heckmann, K., Lindemann, B., Schnakenberg, J.: Biophys. J. 12, 683 (1972).
Hill, T.L.: *Thermodynamics for Chemists and Biologists*. Reading (Mass.), Menlo Park (Calif.), London, Don Mills (Ontario): Addison-Wesley 1968.
Hill, T.L., Chen, Yi-Der: Biophys. J. 11, 685 (1971)
Hodgkin, A.L., Huxley, A.F.: J. Physiol. (Lond.) 117, 500 (1952).
Hodgkin, A.L.: *The Conduction of the Nervous Impulse*. Liverpool: Liverpool University Press 1967.
Hook, C.: Diploma Thesis, Aachen 1975.
Hunding, A.: Biophys. Struct. Mechanism 1, 47 (1974).
Janssen, H.K.: Z. Physik 270, 67 (1974).
Karremann, G.: Bull. Math. Biol. 35, 149 (1973).
Katchalsky, A., Curran, P.F.: *Nonequilibrium Thermodynamics in Biophysics*. Cambridge (Mass.): Harvard University Press 1967.
Laidler, K.J., Bunting, P.S.: *The Chemical Kinetics of Enzyme Action*. Oxford: Clarendon Press 1973.
Landau, L.D., Lifshitz, E.M.: *Statistical Physics*. Course of Theoretical Physics, 2nd Ed., Vol. 5, p. 275 - 277. Oxford, London, Edinburgh, New York, Toronto, Sydney, Paris, Braunschweig: Pergamon Press 1968.
Lasalle, J., Lefshetz, S.: *Stability by Liapunov's Direct Method*. New York: Academic Press 1961.
Lavenda, B.H.: Quart. Rev. Biophys. 5, 429 (1972).
Leiseifer, H.: Diploma Thesis, Aachen 1975.

Lindemann, B.: In: Parson, D.S., Kramer, M. (Eds.): *Intestinal Permeation*. Amsterdam: Excerpta Medica 1976.
Lotka, A.J.: J. phys. Chem. 14, 271 (1910).
Mac Arthur, R.H.: Theor. Pop. Biol. 1, 1 (1970).
May, R.M.: *Stability and Complexity in Model Ecosystems*. Princeton (New Jersey): Princeton University Press 1973.
Mc Ilroy, D.K.: Math. Biosci. 7, 313 (1970 a).
Mc Ilroy, D.K.: Math. Biosci. 8, 109 (1970 b).
Meixner, J.: *Network Theory in its Relation to Thermodynamics*. In: Proceedings of the Symposium on Generalized Networks, p. 13 - 25. New York: Polytechnic Press of the Polytechnic Institute of Brooklyn 1966.
Minorsky, N.: *Nonlinear Oscillations*. Huntington (New York): Robert E. Krieger 1974.
Morowitz, H.J.: *Entropy for Biologists*. New York, London: Academic Press (1970).
Nicolis, G., Portnow, J.: Chem. Rev. 73, 365 (1973).
Noyes, R.M., Field, R.J.: Ann. Rev. phys. Chem. 25, 95 (1974).
Oster, G.F., Perelson, A.S., Katchalsky, A.: Quart. Rev. Biophys. 6, 1 (1973).
Prigogine, I., Defay, R.: *Chemical Thermodynamics*. Longmans, Green and Co.: London and Harlow 1954.
Robinson, P.J., Holbrook, K.A.: *Unimolecular Reactions*. London, New York, Sydney, Toronto: Wiley Interscience 1972.
Sauer, F.,: In: Orloff, J., Berliner, R.W. (Eds.): *Handbook of Physiology, Sec. 8: Renal Physiology*. Washington/D.C.: The American Physiological Society 1973.
Schlögl, F.: Z. Physik 253, 147 (1972).
Schlögl, R.: *Stofftransport durch Membranen*. Darmstadt: Dr. Dietrich Steinkopf 1964.
Schmidt, R.F. (Ed.): *Fundamentals of Neurophysiology*. Berlin, Heidelberg, New York: Springer 1975.
Tellegen, B.D.H.: Philips Res. Rep. 7, 259 (1952).
Thoma, J.U.: *Introduction to Bond Graphs and their Applications*. Oxford, New York, Toronto, Sydney, Paris, Braunschweig: Pergamon Press 1975.
Tyson, J.J.: J. chem. Phys. 58, 3919 (1973).
Tyson, J.J.: J. math. Biol. 1, 311 (1975).
Verhulst, P.F.: Nouv. Mem. Acad. Roy. Bruxelles 18, 1 (1845).
Verhulst, P.F.: Nouv. Mem. Acad. Roy. Bruxelles 20, 1 (1847).
Volterra, V.: J. Conseil Permanent Intern. Exploration Mer III, 1 (1928); translated in: Chapman, R.N.: *Animal Ecology*. New York: McGraw Hill 1931.

Subject Index